Design Participation

Proceedings of the Design Research Society's Conference

Manchester, September 1971

Edited by Nigel Cross

ACADEMY EDITIONS

First published in Great Britain in 1972 by Academy Editions,
7 Holland Street, London W.8.

Text set in 11/12 pt. IBM Press Roman, printed by photolithography, and bound in Great Britain at The Pitman Press, Bath.

CONTENTS

FOREWORD

The unique characteristic of a conference is that it brings together in face-to-face contact a self-selected sample of people interested in a particular field of study or development. No other communications medium offers the opportunity for such immediate and varied personal contact as does a conference. However, many conferences seem to be planned in such a way as to make the initiation of personal contacts particularly difficult. Information on participants is often incomplete and difficult to get hold of; the formal programme tends to dominate the proceedings at the expense of informal contact; facilities for impromptu meetings and discussion groups are often absent; and the conference often seems to be designed for the convenience of the organisers rather than for that of the participants.

This situation is perhaps not over-critical when the conference is held by a mature Society in a well-defined field of interest. However, when neither of these conditions holds, traditional conference design is inadequate. Such was the case for the Conference whose proceedings are reported in this book. It was the first major conference of the Design Research Society, and the topic – User Participation in Design – was expected to (and did) attract people from a wide range of disciplines – most of whom would be strangers to each other. The conference organisers therefore felt it very necessary to make this event an experiment in conference design.

Firstly, we took some care in our choice of venue. We wanted a layout that would help rather than hinder participation and contact-making. Owens Park, the main halls of residence for Manchester University, proved to be ideal in this respect, in that all rooms and facilities were grouped close together. Also, the Manager and his staff were very helpful and responsive to our requirements. For this sort of conference such venue characteristics are vital. The fairly common set-up, with delegates dispersed around a city in hotels, or with accommodation separated by a few miles from lecture halls, would have severely hampered the achievement of the participatory aims of the conference organisation.

We then provided a number of facilities which we hoped would enable participants to make the most of the opportunities for personal contact afforded by their simultaneous presence in Manchester. For example, a participant information system called HOST (Helping Organise Selective Togetherness) was devised, and proved particularly useful in the early stages of the conference for initiating contacts. The pre-planned aspects of the programme were arranged in such a way as to facilitate personal contact, and also to attempt to improve the quality of the discussion of the papers. In the mornings, up to four main-topic papers were presented, each followed by a discussion. In the afternoons, the more specialised papers were divided into topic 'workshops'. No discussion now took place in the main hall, but groups were formed in other rooms around the venue. Thus people not vitally interested in certain topics did not have to sit through long discussions, and those who were interested could get together, identify each other and, hopefully, get somewhere with their discussion.

The proceedings in the main hall were video-taped, and also relayed by CCTV to the lounge. These facilities made it possible for participants to choose their own level of participation. For example there were always several groups of people in the lounge, some watching the proceedings (and making notes in more comfort than in the Hall) and some discussing points made by the speaker. The video recordings were made available to participants if they wanted to see a presentation again, or to catch up on one they had missed. Extra discussion rooms were also in use during each evening. One group took the opportunity to introduce a new participatory design-educational game called GRIPS (Gaming, Random Interfacing and Problem Structuring). Another important impromptu event was a discussion with members of a community action group from Liverpool.

A key feature of the conference was the make-up and attitudes of the secretariat. All were amateurs as far as administration and equipment operation were concerned. Further, most of them were at the conference as participants. The overriding policy of the Secretariat group was to provide any service required of them with a minimum of fuss and no bureaucracy. As a result, participants made full use of equipment, facilities and space available.

There is little doubt that the conference was a success – from the various points of view of the Design Research Society, the organisers and the participants.

What deficiencies there were arose mainly from the fact that all the people involved were new to conference organisation (although this was not always a disadvantage). Finally, the conference was a financial success even though charges were small compared with other conferences of similar size. This has made it possible for the Design Research Society to present a copy of this book to each participant – which to us is the best possible way of saying "thank you" to all those who took part.

R. J. Talbot.
Conference Organiser

on behalf of the Design Research Society
and the Conference Secretariat.

PREFACE

Any activity concerned with changing the man-made world can justifiably be called a design activity. In this respect, most of us are involved in some kind of 'designing' most of the time. But the really crucial areas of decision-making at the interface between technology and society are largely the prerogative of specialist professional designers – engineers, planners, architects and industrial designers. These professions, however, are all currently involved in radical changes affecting their working methods and their relationships with society.

In particular, there is mounting pressure for wider sections of society to participate in the processes of planning and design. This pressure ranges from protest groups fighting undesirable side-effects of technological development, through calls from Government committees for citizen participation in planning, to proposals from designers themselves for adaptable environments which the users may modify directly. User participation, by involving in the design process those who will be affected by its outcome, may provide a means for eliminating many potential problems at their source.

Many designers view the prospect of user participation in design with some concern, while most laymen probably still see design processes as secretive and mystical. To explore some of the possibilities and problems, the Design Research Society sponsored an international conference on 'Design Participation', in September 1971, which brought together a wide range of people whose interests overlap in this area. The end result of the changes under way and reported at the conference may well be to blur the current distinctions between 'designer' and 'user': designing may not always continue to be the exclusive prerogative of professionals.

This book presents the proceedings of the Design Participation conference. The topics covered by contributors to the conference range over a catholic field: social technology, participation in planning, adaptable environments, computer aids and design methods. In each of these topics, the proceedings include leading contributions to the development of the concept of design participation from eminent designers, teachers and researchers from Europe and America.

The contributions have been arranged here as far as possible in related groups, but the book is not divided into sections. I have provided an introductory review of the papers which gives my view of how they relate to each other and between groups. The paper by Reyner Banham was the opening contribution to the conference, and it sets 'participation' in the context both of the design world and of society at large. At the end of the conference, John Page's contribution was a summing-up of the proceedings, and there were also the further closing comments of Chris Jones and Robert Jungk. All the author's references have been collected together, in alphabetical sequence, at the end of the book.

Nigel Cross

The Open University, Buckinghamshire.
April 1972.

ACKNOWLEDGEMENTS

This book is the proceedings of a conference, and therefore would not have been possible without the one hundred and fifty people who variously organised, contributed to and attended that conference.

The idea of holding the conference was born in conversation between Chris Jones and myself, and we were joined early in the planning by Reg Talbot. This three-man committee made itself responsible for the conference on behalf of the Design Research Society, but we were helped as occasion demanded by the officers of DRS – principally Derek Beck, Peter Booker and Doug Hykin. We also had assistance from the institutions we worked for: the Open University and the University of Manchester Institute of Science and Technology. Both Professor Geoffrey Holister (Dean of the Faculty of Technology of the Open University) and Professor Denis Harper (Head of the Department of Building at UMIST) provided generous support and encouragement.

The principal acknowledgement must be to Reg Talbot, however, as it was he who bore the main burden of organising the event in Manchester. He was fortunate in being able to call on a large number of helpers in the days immediately before and during the conference, and we remain indebted to the following people for providing the efficient but relaxed and good-natured organisation of the conference facilities: Andrea Cordani, Shirley Crabtree, Barrie and Daphne Evans, Bob and Sylvia Fowles, Robin Jacques, Ron and Ruth Levy, Owen Lewis, Gladys Moss, Jane Oldfield, Len Warshaw and Richard Williams.

We are also obviously indebted to the contributors, who all provided quite freely the time and thought for preparing their papers, and many of whom travelled thousands of miles to attend the conference. We were very fortunate in being able to obtain so many outstanding contributions.

I would like to express some thanks too, to everyone else who attended the conference, for providing the stimulating atmosphere of the event, and such "high quality listening", as Chris Jones said.

Both before and since the conference, I have been dependent on the secretarial skills of Jenny Holman and Kitty Gleadell. They have, in particular, meticulously prepared the typescript of this book.

Finally, I must thank the staff of Academy Editions, for being such enthusiastic publishers.

Nigel Cross

CONTRIBUTORS

Ignacio Armillas

Assistant Professor,
School of Architecture,
Nova Scotia Technical College,
Halifax,
Nova Scotia,
Canada.

Professor P. Reyner Banham

Professor of History of Architecture,
School of Environmental Studies,
University College London,
Gower Street,
London WC1E 6BT.

Charles M. Eastman

Associate Professor,
Director, Institute of Physical Planning,
School of Urban and Public Affairs,
Carnegie-Mellon University
Schenley Park,
Pittsburgh,
Pennsylvania, 15213,
U.S.A.

Dr. C. R. Evans

Computer Science Division,
Department of Trade and Industry,
National Physical Laboratory,
Teddington,
Middlesex.

Alberto Feo

Design Research Laboratory,
University of Manchester Institute
of Science and Technology,
Sackville Street,
Manchester M60 1QD.

Yona Friedman

42, Bd. Pasteur,
Paris 15,
France.

Professor J. C. Jones

Professor of Design,
Faculty of Technology,
The Open University,
Bletchley,
Buckinghamshire.

Dr. Robert Jungk

Steingasse 31,
A.5020, Salzburg,
Austria.

Dr. P. H. Levin

Lecturer,
London School of Economics and
Political Science,
Houghton Street,
Aldwych,
London WC2.

Professor T. A. Markus

Professor of Building Science,
Department of Architecture and
Building Science,
University of Strathclyde,
George Street,
Glasgow C1.

E. Matchett

Matchett Training,
14 Montrose Avenue,
Redland,
Bristol BS6 6EQ.

Dr. T. W. Maver

Senior Lecturer,
Director, ABACUS,
Department of Architecture and
Building Science,
University of Strathclyde,
George Street,
Glasgow C1.

William J. Mitchell

Assistant Professor,
School of Architecture and
Urban Planning,
University of California,
Los Angeles,
California, 90024,
U.S.A.

Nicholas Negroponte

Assistant Professor,
School of Architecture and Planning,
Massachusetts Institute of Technology,
Cambridge,
Massachusetts, 02139,
U.S.A.

Jeff Nuttall

Senior Lecturer,
Department of Fine Art,
Leeds Polytechnic,
Leeds.

Professor J. K. Page

Professor of Building Science,
Department of Building Science,
University of Sheffield,
Western Bank,
Sheffield S10 2TN

Stephen Platt

Apdo. 6685,
Caracas,
Venezuela.

Dr. R. Roy

Lecturer in Design,
Faculty of Technology,
The Open University,
Bletchley,
Buckinghamshire.

Professor J. N. Siddall

Department of Mechanical Engineering,
McMaster University,
Hamilton,
Ontario,
Canada.

Peter Stringer

Research Fellow,
School of Environmental Studies,
University College London,
Gower Street,
London WC1E 6BT.

Sean Wellesley-Miller

Research Fellow,
School of Architecture and Planning,
Massachusetts Institute of Technology,
Cambridge,
Massachusetts, 02139,
U.S.A.

K. G. Williams

44 Langford Green,
Champion Hill,
London SE5.

For one reason or another, it was not possible to include in this book the contributions made at the conference by Professor D. H. Chaddock, Peter Kamnitzer, Roy Madron, William Osborn and James Taggart.

HERE COMES EVERYMAN

Nigel Cross

For the layman, who is on the receiving end of the planning and design processes, much of what the various professionals hand down to him must seem a very mixed blessing. Every development seems to hold as many threats of harmful side-effects as it holds promises for the enhancement of society. Too frequently, the most that the threatened layman can do is to protest when it is already too late. Not only is he not consulted even about proposed developments in his own neighbourhood, but planning and decision-making at all levels are often deliberately kept secret.

Yet the professional designers in every field have failed in their assumed responsibility to predict and to design-out the adverse side effects of their projects. These harmful side effects can no longer be tolerated, and regarded as inevitable, if we are to survive the future. The increasing amount of protest against a wide range of dubious developments is an indication that many people are now not prepared to go on accepting the rising "price of progress".

A popular response to this conflict has been to call for wider participation in the planning and design processes. There is certainly a need for new approaches to design if we are to arrest the escalating problems of the man-made world, and citizen participation in decision making could possibly provide a necessary reorientation. Hence this conference theme of "user participation in design".

The conference covered a wide range of topics relating to design participation, being concerned not only with conventional approaches to participation in planning, but also with socio-technical issues of who is to control the future, with possibilities of adaptable environments, and with the relevance of the new computer-aided design and design methods fields which could break the existing professional monopolies in design expertise.

In opening the conference proceedings, Reyner Banham raised this very question of professional expertise, and put his finger on an issue obviously of central concern to the conference: the concept of professionalism. The professional man, who is trained to solve a particular type of problem in a particular way, is living on the knife-edge of a paradox. If he is truly successful in solving his particular type of problem, and removes the root cause from which the problem arises, then he also destroys his own livelihood. We can never really trust a professional man, Banham implies, because he inevitably has a vested interest in his own type of problem continuing to exist. This is an uncomfortable thought, which must come home to roost with many of us.

The conclusion which Banham reaches is that only by breaking through "the rules of the game" – the many games established and run by the professions – can we hope to approach a de-professionalised future of any significance. The 'Alternative Culture' is showing the way, with its relaxed, un-polarised attitude to a game of life which has no fixed rules. 'Participation', in any radical sense, is about giving all the people access to the tools, resources and power which have been the jealously-guarded prerogatives of the professionals.

SOCIAL TECHNOLOGY

We are, or have been, prisoners of the technology of our time. Professionalism is a particular kind of specialisation, and specialisation – the division of labour – is the technique of production-line technology. As we develop new technologies we will develop new roles and new images of ourselves.

Jeff Nuttall suggested that professionals have been prone to adopt simplistic images of their clients. But perhaps this is an inevitable facet of specialisation, which continually requires the few to construct a small range of models of the many. The Modern Movement in design has failed simply because of this. It has assumed the ethos of functionalism, that a person's needs can be defined and modelled and translated into objective artifacts which satisfy those needs. To define a person's needs, Nuttall says, is to deny that person's humanity, for the essence of being human is to be undefinable, to retain one's mystery. We need now a technology conducive to the mysteries of life – ecstacy, love, pleasure and excitement.

An artist's views of the need for alternative technologies, however, is unlikely to match the views of the technocrats who have the power to choose the future. Robin Roy's analysis of the control of the future articulates a growing doubt that the future should merely be "a larger and glossier version of the present", and he is able to list a number of developments which hint at a possible shift of control. But he also sees that any radical shift will meet great resistance from the establishment. Only a multitude of small actions – many of which may appear fruitless at the time – will generate enough strain in society for real change to occur.

Peter Stringer's paper offers some careful thoughts on the meanings of 'design participation'. He suggests that participation can mean variously having or doing or being a part, and he perceives that there may, in fact, be a general progression in attitudes through this spectrum of meanings under way in our society. Stringer relates our current concern with individuality, change and a more personal control of the future to a set of philosophical axioms – the basis of the construct systems which enable us to make sense of the world.

Although they each express it in very different ways, there seems to be a common theme between Nuttall, Roy and Stringer. Technology, society and man's view of the world are inextricably linked. Trying to change any one of these must inevitably go hand-in-hand with changing them all.

PARTICIPATION IN PLANNING

Probably the most well-known concept of design participation is that of a wider participation in urban and regional planning. Representing what may perhaps be regarded as a now conventional approach to participation in planning, Peter Levin develops a model of the planning process based on its component administrative, technical and political processes. He points out that a planning decision is "an act of choice which generates commitment to a specified course of action", and that there are thus two important properties to a chosen course of action: a) it possesses specificity, and b) commitment is attached to it. The more specific a course of action is, and the more commitment that is attached to it by its promoters, the less likely is it to be modified under attack. Therefore, those who wish to participate in planning decisions must seek involvement before any course of action becomes too specific or has too much commitment attached to it.

Levin explains how each of the three sub-processes – administrative, technical and political – tends to raise specificity and commitment, and he suggests that only in the political process – being the only one to involve interaction between different groups – is participation possible. This means that the administrative and technical processes should be formulated in such a way that they do not wholly determine to which specification commitment is attached, nor the total amount of commitment generated, but leave the maximum scope in these respects to the political process. Levin concludes with some advice for those who, as is becoming common, find that they need to force their way into the planning process in order to block some development which would adversely affect them.

A relatively new technique which may promote wider participation in planning is the use of planning 'games'. Originally developed as educational exercises for planners, these games enable the players to explore roles and conflicts in resolving planning issues. There is an obvious potential for incorporating representatives of the user population in the role-playing groups, and Ignacio Armillas describes a game which has been developed in this way. 'URBANISTA' is a gaming exercise for both designers and users.

Alberto Feo's approach to the gaming technique is even more specifically oriented towards user involvement. He suggests that the various protest groups coming into existence can be seen as the embryonic development of a new socio-technical control process which should be encouraged. His application of operational games has been towards the development and evaluation of alternative strategies for these protest groups.

A doubt which remains, however, is whether the games tend to reinforce or to weaken the conventional planning roles represented by the players. The playing of roles in real life – i.e. being forced to narrow oneself into a specialism – is, as we have discussed, probably a fundamental current socio-technical problem.

ADAPTABLE ENVIRONMENTS

The traditional planning and design processes may well become obsolete if the proposals for continuously-changing, do-it-yourself, adaptable environments become reality. There has been a number of these proposals from the architectural profession in recent years. (The fact that the proposals come from the professionals, rather than from the users, should perhaps arouse some suspicion – remember Banham's warning about the expert's interest in his problem continuing to exist.)

Yona Friedman proposes removing the professional designer from the design process, by providing instead an appropriate 'repertory' of environmental technology, and a 'warning' feedback mechanism to keep everyone informed of the consequences to the whole community of each individual choice within the repertory. The professional designer is not altogether eliminated; he adopts the role of a technician in preparing the repertory.

Friedman's analysis of the need for this professional change of role echoes Jeff Nuttall's criticism of the de-humanising aspects of 'functionalism'. Because the professional designer has to attempt to satisfy the needs of a large number of users, he is forced to model his perception of needs on a hypothetical 'average' user. This in itself is sufficient reason for replacing the existing paternalistic design processes by a neutral 'infrastructure' (if any technology can ever be neutral), but Friedman's most damning criticism of the professional design process is that it separates decision-making from risk-taking – the designer makes the decisions, but the user takes the risks.

Charles Eastman's proposal for an adaptive-conditional architecture is based on the need to achieve a measure of fit between activity and environment despite three principal difficulties: designing for anonymous users, designing for unpredictable behaviour in new environments, and designing for activity patterns which change over time. Adaptive-conditional architecture would offer total environmental control which the user could regulate to individual requirements. Eastman identifies some architectural trends towards this more personalised environment.

Similarly, Sean Wellesley-Miller argues for a tactical, on-line (i.e. piecemeal, directly user-controlled) design process to replace the current strategic, off-line (i.e. comprehensive, remotely designer-controlled) process. Perhaps it is not surprising that his examples of actual on-line, tactical design in practice are drawn from societies based on primitive technologies. The new technologies we have been discussing may generate a society which has much in common with pre-industrial societies.

COMPUTER AIDS

Many of the proposals for adaptable environments assume the existence of sophisticated computer installations for providing monitoring, controlling and up-dating functions. Nicholas Negroponte's paper explores the bridges necessary between environmental hardware and computing software to achieve a 'responsive' architecture.

The Architecture Machine Group at MIT has concerned itself with aspects of artificial intelligence in the context of architectural design (hence 'the architecture machine'), and Negroponte offers his paper here as a first step by the Group towards extending this work into researching the intelligent environment. He identifies three aspects of intelligence which the environment must possess – recognising, responding and learning – and discusses alternative examples of achieving these through computation. Obviously, these examples raise more questions than they provide answers, and Negroponte is the first to admit that he only yet has inklings of what living with responsive architecture would be like.

On the face of it, a far less radical computer application is Christopher Evans' computer-patient interaction for medical diagnosis. Yet, translated directly into an environmental context, this application would probably meet much greater resistance from the professionals than Negroponte's proposals do, because it is a much more immediate and comprehensible de-professionalising act. What Evans has done is to enable a machine to perform what was hitherto regarded as a professional art. Evans' medical diagnosis program is, as yet, still kept under a professional thumb – it passes the data it collects from the patient on to a doctor for final diagnosis. But it seems clear that the machine could as well make the diagnosis, once we lose our fear of mechanical fallibility, or could return the data to the patient together with the rules of diagnosis, for the patient to make his own diagnosis. It may be stretching the meaning of the word to call this 'design' but it is clearly a fundamental example of 'participation'.

Although there are not, as yet, any such clear-cut applications of computer aids to participation in the design field, there are many examples of recent developments in computer-aided design which could be given a participatory twist. William Mitchell reviews a number of such developments, and discusses their relevance to participation. A specific proposal for a mechanism of computer-aided design participation is made by Tom Maver. He suggests the incorporation of a 'solution team' – composed of client, users and others affected by the design project – in a cyclical design process in which the designer's proposals are submitted to appraisal by computer.

The potential of Michell's and Maver's examples for de-professionalising the art of design becomes manifest in the light of Evans' example of de-professionalising the art of medicine. Current experiments with computer aids will doubtless have profound consequences in many professions, and help shift the decision-making back to the risk-takers.

DESIGN METHODS

Computer-aided design techniques have been one major aspect of the more general study of design methods which emerged in the 'sixties'. Now that design methodology has become established as an academically respectable subject, however, some of its hitherto leading promoters have begun to turn and express doubts. One reason may be that, as Tom Markus pointed out, although many design models have been produced, none appears to recognize the established social and political status of the designer. Hence the models, and the methods derived from them, become irrelevant if one wishes to consider the wider socio-political context of design.

Markus perhaps comes to the crux of the issue in discussing the possible attitudes which designers could now assume. He suggests that there are three main alternatives: 1) to promote expert professionalism, which depends on stable social structures, legal protections, etc., and continue to function only by patronage from the centres of power; 2) to adopt a sympathetic stance to design participation, accepting the growth of professional bodies to encompass new disciplines, etc., and adopting new means for developing ranges of possibilities for public choice, which should effect some compromise between the planners and the planned; 3) to reject both these previous possibilities and work for "a real transfer of power on design decisions", generally through unpaid work with groups such as tenants in twilight housing, factory workers or hospital patients, and adopting a midwifely role. This spectrum of roles implies that a concern with design methods and computer aids for 'participatory' design could merely be a liberal reforming ploy; the revolutionary will not be satisfied until design control is truly liberated from professionalism.

We can now see more clearly the attitudes underlying the contrast between James Siddall's cooly rational attempt to incorporate user's value systems into the design process, and Jeff Nuttall's polemical concern for a technology which does not violate the 'mystery' of each individual user. Siddall is in the liberal reformist tradition; Nuttall in the radical activist tradition.

There is a similar contrast between Matchett and Williams attempting to liberalise the design of health care facilities, and Stephen Platt getting into the nitty-gritty of the disabled person's life. Matchett and Williams actually raise the question of "who is the user" of the health care system – the patients or the medical professionals? They are concerned to establish a design procedure which will accommodate a wide range of participants, whoever they may be.

By contrast, Steve Platt goes straight to the people with the obvious problem. His research shows up the reality that, where people are forced hard up against the environment, we are nowhere near achieving participation. Appropriately enough for the end of the conference, his contribution stood the concept of design participation on its head. Platt's "design method" involves him participating in the lives of the users, rather than them participating in a design process.

CONCLUDING

John Page was able to produce a remarkably comprehensive summing-up of the proceedings for the closing session of the conference. He also brought in some discussion of political questions which had been largely ignored. This omission of political aspects had been mostly deliberate; in setting up the conference, I was looking for examples of new technologies and new techniques which might be sidestepping conventional political controls. That the conference seemingly *had* to come round to discussing politics, suggests that this may have been a fallacy. But perhaps it only means that I wasn't looking in quite the right places. Like Reyner Banham, I find a small ray of hope still in the Alternative Culture, because, as he said, it *has* managed to avoid the major political polarisations.

Both Chris Jones and Robert Jungk echo aspects of Banham's introduction in their closing comments. Chris Jones refers to the "frightening simplicity" of the professional roles we play, and Robert Jungk points to the break in continuity between generations which we are experiencing. Jungk also offers us another rather frightening vision; of the need for an 'underground' which could survive the coming crises of the technocratic period, to emerge after we have suffered the consequences of "the lack of foresight of our fathers and grandfathers".

I could not really conclude this review without commenting on the conference as an event in itself. Robert Jungk refers to the different atmosphere he found on coming to this conference from another one dominated by an older generation. Everyone at the conference seems to have found it a stimulating event. I take this to be principally due to the informal, neutral technological infra-structure of facilities provided as an attempt to liberate everyone from the production-line of the normal conference.

Finally, I should add that the phrase 'design participation', which I thought I had invented specifically as a title for the conference only nine months earlier, had already become, according to one reviewer soon after the conference, "an inadequate cliché". You have been warned!

ALTERNATIVE NETWORKS FOR THE ALTERNATIVE CULTURE?

Reyner Banham

When one looks down the list of speakers at this conference, and the titles of their papers, one wonders whether we have not got the same old Design Conference, but with the new wonder ingredient 'participation'. It is very difficult not to get that impression – there are all the same old names, beginning with Reyner-bloody-Banham, with the same old part-worn titles, and probably with the same old footnotes, etc., etc. What the hell do we think we're doing?

But the fact is that the wonder ingredient 'participation' hasn't actually been around all that long. If I stand on my own professional skill as an Historian of Contemporary Affairs in the world of Architecture and Design, I only have to go back to 1965, to the Vienna conference of the International Council of Societies of Industrial Design, to recall a situation in which the concept was still unknown. Among the papers at that conference was an early version of Julian Beinart's famous one on the painted houses in the western native township outside Johannesburg. That is to say, that well-known benevolent South African government had given all the coloured workers these neat, efficient, miniscule houses without services and without external finishes, and the enterprising inhabitants had done over the outside with patterns based on Gillette razor blades, rising suns, peanut ads. and things like that, and had made a great contribution to a piece of urban design.

Now for most of the heavy professionals present at ICSID, this idea was quite profoundly subversive and shocking. The idea that there could still be something left in a design for the ordinary consumer to do, was to them a dereliction of duty and a lot of worse things than that – an abandonment of basic cultural standards and all that kind of thing. The shock waves were not loud, but they could be felt. You could feel the shock waves coming back at Beinart, who had apparently offered to kick out one of the legs on which serious community design stood; that the designer would do it all was, I think, the automatic assumption of pretty well everybody who had come to that conference.

That was only six years ago. This past summer, at Alvin Boyarsky's 'Summer Session' at the Architectural Association, two delegates from Japan gave a talk with some-such title as 'Participative Planning in the Tokyo Bay Area', and all the radicals and Maoists went along to hear what was clearly going to be an extra-groovy talk, because it was about something in Japan as well as about 'participation'. But only to discover, to their more-or-less prefabricated horror, that the participants in this participative planning were major land owners around the shore of Tokyo Bay and large development companies; 'the people' were not involved at all. In six short years a concept which had been non-existent had already been narrowed down to a point where it was almost useless. It had gone from something which was absolutely spit-new to all those who heard it for the first time, to a condition where everybody thought they knew what it meant, and were astonished to be reminded that the word can carry a broader spectrum of meaning than they had had in mind. Later, we read David Eversley, no less, proposing in *The Guardian*, that it should become a compulsory part of the education of planners to be sent out into the streets for a year, to be participated at by 'the people'.

Now when one gets to the David Eversley, compulsory training level, one begins to have the feeling that this is, in Donald Schon's terms, one of those 'ideas in good currency' and therefore dead; one of those ideas that everybody has heard of, everybody can discuss, everybody knows what it means. It has reached the point where it is susceptible to government action and has therefore ceased to be a live issue.

But the presence of 150 souls at this conference is a fair indication that it is not quite a dead issue yet. The fact that we think that it is a live issue I take to be important, and something on which, with luck, one ought to be able to operate. If we think it is alive, what are our motives for doing so? Why do we want 'the people' – that convenient abstraction – to participate in the processes of design, whether it be at the commodity level or the community level? In other words, not to put too fine an edge on it, what's in it for us?

Right, so some of us are putting our social consciences to work. We believe, for social, political, religious reasons, that these things should be done. On the other hand, it is not too difficult to see that, in some places and at certain times over the last six years, the motivation of professionals to stir up the populace into participatory action has been a way of finding allies for our own private inter-professional guerilla wars. One of the reasons architects want the ordinary people to participate is a way of getting back at planners, and I can think of examples all down the line. We have looked round for allies and discovered that God-given, God-pure, beyond-criticism ally, the people-out-there.

Our reasons for wanting to get into this scene are extremely mixed, and I think we ought to face that.

Nevertheless, clean or dirty as our motives may be, the fact is that nearly all the operating professions in the field of design, planning and community services of various sorts are in some disarray. We are having a crisis of professionalism. The professions that have served Western society for better or worse since the Renaissance, or even longer, are in a bad fix at the moment. Part of the trouble of course lies within the concept of professionalism itself. Why do we want amateur assistance in replanning our cities, in designing our products? The answer is because we are not at all certain what we are about and how we should be about it.

Professionalism is a very funny thing. It doesn't get, to my mind, the kind of analysis and discussion, including psycho-analysis, that it really deserves. Professionalism is a way of organising competence, of getting certain specialist skills together in a place and in a condition where they can aid the rest of society. But it achieves these specialist skills precisely by specialising, by narrowing its vision, by concentrating on a few things, by practising one thing instead of everything.

The professional is in every sense the opposite not only of the amateur, but of universal man and all those other great educational ideals as well. More than that, a professional is a problem-oriented man. You don't have an expert, it has been said, until you have a problem. No-one is expert on anything until the problem is perceived to exist. And as an expert, he has, I am afraid, a political or financial interest in not seeing the problem finally solved, because he will have done himself out of a job. If he clears up the problem he has been asked to clear up, he is out of work. It is not a light thought, but we are all in this fix. If everybody knew about history, I would be out of a job.

A professional is a man with an interest, a continuing interest, in the existence of a problem. A good doctor is really one you never need to consult; which may be great for you, but it's terrible for the doctor. He needs to be consulted, otherwise, even under the National Health Service, he doesn't get paid. Worse than that, he doesn't get the reassurance that he is doing a good job in the world. He needs to see patients, even if they are perfectly fit, just to persuade himself that they are fit because of his efforts. (That may or may not be true, but it's good for him.) He exists in terms of a problem. Without that problem, without patients, without an expressed, perceived human need out there somewhere, he has got nothing to do and no real reason to exist.

I am an expert on perceived needs right now because we have got a baby in the house. The baby has needs, which it expresses in a standard format. Whatever the needs may be, he makes more or less the same noise. His mother, my daughter, rushes over, analyses the situation with professional skill, decides that he needs a clean nappy or that he needs a drink, that he needs teddy picked up or that he needs moving out of the sun, and acts accordingly. In about 75% of cases she is right – the noise stops – which I think is par for the course for most professional experts. But she is interpreting a perceived need – not answering a formulated need. The baby may not even perceive the need; he may be making the noises by just pure reflex, I don't know. His mother perceives the need and decides what is to be done about it.

There is a perceived need at the moment, in the disarray of the professions, for 'the people' to speak and to be heard. 'The people' themselves may only feel discomfort, pain, disorientation, or something. They may not themselves know that anything can or should be done about it. They believe that their views are in some way not getting through, and they have summoned up already a new class of professionals – the professionals of Vox Pop, who appoint themselves, for the usual array of mixed motives, as the mouthpieces of the people. Traditionally they have been Populist politicians, Trade Union leaders, etc. More recently, men like George Clark and Ralph Nader have set themselves up as people who are skilled in expressing what they perceive to be the needs of the inhabitants of Notting Hill Gate or wherever, or of consumers in general.

But these spokesmen are themselves professional experts. They come from the professional classes, they have a professional background and training. They may have, in some cases, an interest in the problem itself; in the problem continuing to exist. This is maybe no more than a suspicion, but it is a suspicion we ought all to entertain about ourselves.

The worst thing about having a stake in the problem is that it usually means that, in a sense, you accept that problem as the world which is relevant to the argument. You tend to get shut up within the problem itself. You accept the rules of the game. If, for example, you are a consumerist, then, by implication, you accept that the world is divided into producers and consumers. And this can lead to dangerous non-thinking.

Getting shut up within the given rules of the game is dangerous because once both sides have accepted the rules of the game, the game is no longer worth playing. The original objective of the game disappears once both sides know the rules.

Pretty well everybody of my generation and background in England will have grown up with one of the most spectacular cases of this, that is to say the destruction of the working-class intelligentsia by the Workers' Education Association. When W.E.A. was young, one of its main functions was to help the workers to survive and prosper through exploiting the machinery of a bourgeois democracy. It was the never-acknowledged aim of the W.E.A. to equip its members with bourgeois know-how, with middle class responses, so they knew how to play the game in the council chamber, the court room or wherever else – and it succeeded. There is no point in the W.E.A. people complaining that, now, their classes are only interested in art appreciation, flower arranging, music and cissy non-political stuff like that. They themselves had helped to create the market for that kind of classes by successfully helping to destroy, as Richard Hoggart and everybody else has pointed out, the 19th century culture of the working classes. I am not saying that that necessarily was a culture which should have been preserved. Having grown up in the tail end of it I have no grounds for being sentimental about it, but nevertheless its destruction was the inevitable product of the process which the founders of the W.E.A. had put in hand, wittingly or unwittingly. In the end, the particular game of educating the workers had ceased to be meaningful and ceased to be worth playing.

There is something similar which I have had a sideline view of more recently, and that was the Los Angeles goals programme – an attempt to get town planning research to an on-the-street, shop-front level. The planners of L.A., after Calvin Hamilton took over, were going to really find out what kind of city the people of that city wanted. They were going to go into the streets, into the ghettoes, into the suburbs and on the beaches and get the word from the consumers of planning themselves. The people were delighted, and they came up with a long list of complaints and things which they thought needed to be done; "How can I find a parking space downtown?" "How can I get the Mexicans out of my neighbourhood?" All kinds of live issues like that, which of course, for good liberal professional reasons, don't appear in the planner's vocabulary at all – especially the bit about how to get the Mex out of the neighbourhood! So there was a kind of impasse which, when I was first in L.A. in 1965, you could still feel. Then, of course, the planners saw that the thing to do was to explain to the people what planning was really about, and out came these handy little booklets explaining high and low density, high rise and low rise, cluster and distributed. But the people just walked away. The book just didn't tell you how to get the Mex out of your neighbourhood, or how to find a parking lot outside a shop downtown. Once the rules of the game were known to both sides it was seen to be no game. Like, bad thinking stopped play.

Now this kind of situation is, I think, where alternative cultures come in. The concept of an alternative culture is, of course, being heavily hammered. To a great number of people it is no alternative – at least, to a great number of Marxists it is no alternative – because it accepts some of the capitalists' rules of the game. There are such things as underground entrepreneurs, and no-one in the underground seems to mind them making a mild profit. It is the sort of profit that the medieval church might have approved of, rather than the sort of profit that the City looks for nowadays, but nevertheless no-one in the underground minds people actually turning slightly more than the honest penny out of providing sound equipment, lights, places to have pop festivals, and things like that. But the mere fact that this is done, that there is no out-and-out rejection of capitalist methods, means that for a committed body of Marxist thinkers the alternative culture is no alternative.

Now, to me, it is this very thing which makes it an alternative. The rules of the game as between Marxism and Capitalism are defined and are known and are – heaven help us – nearly 150 years old now. It is one of the best known rule books in the business, and anybody can play. Anybody with a normal education could put up a convincing imitation of a Marxist or a Capitalist, whether he was brought up in North America or in Russia, because everybody knows what kind of noises to make, and so on. The function and interest of the Alternative Culture (capital A, capital C for the moment, as far as I am concerned), was that it proposed a third way, containing elements of the other two, with value systems which didn't really belong to either of the other two. It broke a given polarisation. It may not have been a very big break, but it broke a given polarisation which had ceased to be productive in human terms, and it proposed, so to speak, a third term. (This all sounds like quite good Hegelian dialectic to me but that's not how I arrived at these conclusions, though I just wish a few more Marxists might notice.)

Consider attitudes to technology. There are two polarised attitudes to technology, and we all occupy one or the other at different times of the day, I suspect. When we need a hot bath, technology is good; when we turn on the tap, and actual hot H_2O comes out of it, then at that time there is nothing wrong with technology at all. Then we see smoke pouring out of the power stations which are providing electricity to heat the water, and technology is bad. Culturally, we polarise the two extremes: on the one hand, technology is the great provider that gives us the necessary goods to support the standard of life which we expect, on the other hand technology is the great despoiler which is crippling the underdeveloped nations, polluting our own atmosphere, making the ecosystem uninhabitable, etc., etc. We tend to polarise in debate into technologists on the one hand, and beautiful Neo-Luddites on the other.

That is, until you read the *Whole Earth Catalogue*, when you discover that there are *some* beautiful people who can use technology. They use the bit that makes sense to them, the bit that comes to hand. I know the phrase "soft-centre, New Mexico technology" is meant as an insult in some quarters, but up there in New Mexico they are using the technology that comes to hand. They are sawing the roofs off cars to make domes and things, they are making beautiful sculpture out of abandoned driving mirrors, they use the mule to plough one field and the tractor to plough the other. It is a very relaxed attitude.

It is an unpolarised attitude, and what it is all about really is that there are no rules to their game. What makes it so relaxed, what makes it unpolarised, what makes it for me a small ray of hope, is that it ignores the supposed rules of 'bad technology' and 'good technology', and says instead, "Put the hardware in our hands and we will invent the rules."

"But", says a small frightened voice inside all of us, thinking about 'Doctor Strangelove', and so on, "technology is sort of complex and dangerous and stuff." All right, complex it certainly is – I mean it is so complex that you can make LSD in the bath tub. The average household, certainly the average middle class professional household, is already equipped with the technology to make Acid. If you can develop films in your bathroom, you can make LSD in your bathroom. The Alternative Culture has proved this to be a workable, commercial and psychedelic proposition.

Electronic technology is so complex they failed to teach it to me at school – they didn't try awfully hard, I must admit, but they didn't get anywhere with it at all. I didn't notice anybody teaching my kids, but they can handle most of the electronic gear that exists. The whole concept of community TV, which I dearly love, is a standing demonstration that there is no technology too complex for almost anybody to use it – as long as it comes in reasonable-sized packages.

I am not saying anybody can run a blast furnace, though I am not sure whether anything so antediluvian as a blast furnace ought to exist, even within the technology of metals, these days. But if you can pick it up and carry it about, or put it in the back of a Transit, there is hardly a piece of technology around that an average intelligent person can't master.

Now this is pretty interesting stuff I think, and its social consequences and its political consequences aren't really being got at yet. I am certainly neither sociologist nor politician enough myself to do more than to divine what appears to be an interesting looking situation.

It is what the *Whole Earth Catalogue* is all about – where to find the resources to do what you want to do, with your own set of rules. The indication which I deduce from this, is that do-it-yourself is the only real design participation. When the resources are in the hands of (here they come again!) 'the people', and 'the people' invent their own rules for the game, then I think design participation is getting somewhere.

HOW TO USE TECHNOLOGY

Jeff Nuttall

We are at a time of failing confidence. The explosion of heroic egos on whom we still feed – the vision of Le Corbusier, Frank Lloyd Wright, Walter Gropius, and of the Cubist and Non-Figurative painters who inspired them – can now be seen in retrospect for the first time, and we are filled with questions arising out of their sociological failure. In our progress towards these failures we have maintained a number of values as being absolute; hygiene, equality and truth.

In the early years of this century two statements were made which can be held in opposition to one another. The first is Gertrude Stein's "A rose is a rose is a rose" which as she later pointed out implies what Picasso had discovered, that "A painting is a painting is a painting". The second statement is René Magritte's picture of a tobacco pipe bearing the inscription "Ce n'est pas une pipe".

Design has followed the first of these. We believe that a building is a building is a building, and a chair is a chair is a chair. It has espoused the principles of reality and definition just as it has adopted the geometric forms that were pioneered by Gertrude Stein's painter friends. Thus design has championed truth-to-material rather than fantasy, democratic uniformity rather than variety and individuality, and hygiene rather than comfort. Like doctors and priests, architects have carried a series of very clear pictures of what and how people ought to be.

Hygiene, equality and truth have, then, begotten their own negative qualities; prudery, anonymity and spiritual sterility. The manifestation of these ills in sexual neurosis, psychosis and crime is the typical social difficulty that leads us to re-examine our design orientation. We are in a world of massive urban misery. We are making the wrong goods the wrong shape, and all that I, as a poet, can do is to try to suggest why this is so and what we can do about it. All human beings are different. We can't say a man is a man is a man. We have to say a man is Charlie Jones, Bill Smith, Aloitius Carberry, Meredith Williams, and so on. The things men share in common are of course vital, so vital they are still unmentionable in many public places. They share their gastric and sexual functions. So vital are these functions that we are traumatic about them and have cloaked them in multiple layers of taboo and fear, so as to invest them, the least mysterious of activities, with all the qualities of *metaphysical* existence. For beyond the physical, men share the quality of each and every one being completely different. Along with his alimentary system and his genitals each man has his unknown factor, his mystery, in which he can play perpetually with the different undefinable subjective toys called love, vision, God, art, beauty, good; turning and inverting those bright magic balls as ideas change and society follows suit.

The way in which ideas change, however, is the result of two conflicting processes. One is by the inventions of single influential figures of genius, and the other is by an organic accumulation of customs and artifacts arising out of the *incalculably varied* behaviour of men.

The force of genius is vital to art and vital in the precipitation of those existential crises in one of which we are currently caught; whose purpose is to enliven the faculties of man and expand what might otherwise be a very mundane field of experience. And this is best done by art and outrage, the proper field of genius.

The other process is that whereby the mysterious complex of human activities manifests its aesthetic and its direction over which no individual holds control or even complete understanding. Because the second mode is constructed from, indeed is an important thread in, the fabric of total human activity, it is in this second mode that total human needs are best accommodated. For high art knows nothing of need and will deal with it crudely in the interests of creative freedom. Or it will postulate Functionalism which is an attempt to polarise the physical necessity from the metaphysical desire and thus relegate the people to a state defined by mere need, and reserve the luxuries of imagination for the privileged genius.

We are designers and if we love the people we design for, and love them well, we must address ourselves to the second manner of creativity. Our role is subtler and more passive, more telepathic and communal than that of genius. What we have to do is to accommodate the infinitely variable. We can see that we, like doctors, sociologists and politicians have defined too glibly and too hastily what is inherently undefinable.

But the question arises; what else can a designer do but act according to his own ideas, and his own ideas can surely be nothing more than a definition of human needs, for better or worse?

Now I don't think this is necessarily the case because I don't think it is the only way a designer can work. We are living at a time when our spirits are being salvaged at the last moment by the aesthetic of the *objet trouvé*, by collage and assemblage. Not only has the sterility of non-figurative painting and sculpture been relieved by the talents of men like Keinholz, Beuys, Muehl, but the slums of surviving 19th century cities, the shanty towns, the visual chaos of uncontrolled advertising and shop fronts, in other words the areas of urban collage, the remnants of former societies, because they constitute artifacts which are accumulations and therefore express the total psyche of the community, are proving richer and less damaging places, where people are happier, than the cleanliness and space of housing estates and architect-planned towns. I put it to you that there is a substantial difference in kind between the ills of poverty and overcrowding and the ills of spiritual sterility and alienation. When you consider alcoholism, robbery with violence, malnutrition, and disease, these slum evils, terrible though they may be, are of a lesser order than the evils of boredom, bourgeois hypocrisy, the vicious structure of psychic brutality which is called Respectability, aimless psychopathy of the skin-head kind, drug addiction even amongst children, psychosis and suicide.

When your mystery has been categorised and catered for according to another man's definition of you, even though that man is a genius, you are threatened with psychic annihilation unless you protect yourself by your only possible means, the violation of your own personality insofar as it has become an artificial role.

A person's mystery requires a margin of uncertainty in its environment whereby it can retain its own ambiguity, and indulgence of these ambiguities is called adventure and discovery; it is the source of animation, the prerequisite of joy and delight.

That many of these things were recognised by Frank Lloyd Wright does not answer the problem. Lloyd Wright's enthusiasm for intuitive form did not result in intuitive architecture. It resulted in a pastiche of intuitive architecture by a brilliant and self-conscious artist, an idea of intuition imposed on the intuitions of man.

People may get their high idealism from high art and they must, I believe, get their energy from high art, in their own good time. But meanwhile designers have to provide an urban environment which, like a good lover, is varied, unpredictable, ready for anything, and not afraid of a bit of violence or a bit of dirt. To provide this we are going to have to join the people in a situation of mutual improvisation. We should look at what people currently make for themselves – the Facteur Cheval's garden, the shanty towns, the allotment sheds, the drop cities, the strange encampments that spring up at pop festivals – and we should sympathetically read and conjoin the aesthetic of these maquettes. Having done that we should redefine ourselves as technical advisers, providing a vast and subtle range of methods, so that under the instructions of the prospective user, houses and goods can re-inforce and amplify the idiosyncracies and fantasies of that individual. Thus a community might construct its own lasting monument, ritual effigies, primitive ancestors to provide a magical sense of identity, as opposed to a rational, authoritarian one, for subsequent generations. One would get cities as wildly imaginative as the Watts Towers or Californian custom cars, as rocker jackets or railway topiary, as patched jeans or Medieval carving, as lavatory doors or cave paintings.

Now our technology is ill adapted to this because not only have we allowed scientists to create for the human being according to their own lamentable definition of what a human being should be, but also we have subscribed to the idiotic theory that technology possesses a wisdom and power all its own, and if our technology conducts us over the edge of a cliff all we can do is prepare for the fall. We none of us particularly want to travel faster than we can think, hurt more human beings more viciously and more systematically, be bored literally out of our minds, poison the air, listen to Malcolm Muggeridge's moving photograph, go to work on a series of lifts and escalators, or go to the moon. What we want from life is ecstacy, love, pleasure and excitement and the energy to achieve it. A technology that doesn't provide an environment conducive to these things is, in my view, no technology at all. For a technology is a technique and no technique can function as such unless we first know, in every subtle shade of our human awareness, what the hell we are going to use it for.

CHOOSING THE FUTURE

Robin Roy

In industrial nations, the power of choosing virtually everything from leisure facilities to the type of food we eat, has passed into the hands of specialist experts. These experts – urban planners, management consultants, sociologists, market researchers, systems analysts, economists and so on – advise decision-makers in government and business who have the power to translate this advice into new products and systems – housing estates, cars, weapons, schools, foods, aircraft, welfare services, and all the other components of industrial society.

Societies in which the control of affairs is governed largely by appeal to specialised technical expertise have been called 'technocracies'. The many recent attacks on technocracy appear to focus upon three issues related to choosing the future. These are:
1) that technology is out of control and we are caught in the spiral of 'progress';
2) that because of the complexity of the technical arguments involved, the ordinary citizen cannot participate in decision-making;
3) that the technocracy offers a single vision of the future, namely a larger and shinier version of the present.

In an influential and highly pessimistic commentary on the technocracy, Ellul (1964) has said:

> "The principal law of our age is that everything which is technique is necessarily used as soon as it is available without distinction of good or evil. For example the atomic bomb is a necessary stage in the evolution of the technique of harnessing atomic energy."

This is perhaps the extreme view implying that technology is an unstoppable force with a life of its own. One reason for the apparent uncontrollability of technology is offered by Rose and Rose (1970):

> "The absence of effective political control means that scientific developments are allowed to proceed from the point where they are nothing but a gleam in a research director's eye to that at which they are so technically sweet that they are virtually impossible to rescind without their ever being subject to public scrutiny. The technological imperative will have driven them forward with the public only knowing what has been done in its name post hoc. The success stories and the scandals are launched upon a society, its politicians and most of its scientists without the possibility of passing judgement upon them in time. When did the community decide that it wanted to invest money and skills into the development of heart transplants? Or supersonic airliners? Or chemical and biological warfare?"

This leads us to the second major criticism of the technocracy – the lack of opportunity that the individual citizen has for participation in making decisions. In Theodore Roszak's impassioned attack on technocratic society he says (Roszak, 1970):

> "In the technocracy nothing is any longer small or simple or readily apparent to the non-technical man. Instead the scale and intricacy of all human activities – political, economic, cultural – transcends the competence of the amateurish citizen and inexorably demands the attention of specially trained experts. In the technocracy everything aspires to become purely technical, the subject of professional attention. The technocracy is therefore the regime of experts – or those who can employ them. Among its key institutions we find the 'think-tank', in which is housed a multi-million dollar brainstorming industry that seeks to anticipate and integrate into the social planning quite simply everything on the scene. Then, even before the general public is fully aware of new developments, the technocracy has doped them out and laid its plans for adopting or rejecting, promoting or disparaging."

The key issue here I consider is not so much Roszak's view that technology is incapable of serving human needs and that the scientific world-view is necessarily inhuman, but that in the technocracy decisions which will determine to a great extent the way millions of people will live their lives are made in semi-secrecy by a relatively small number of experts and planners. The public is rarely even consulted about the desirability of these choices (except perhaps by market research or social survey without their realising the implications), and often the individual is unaware of the choices being made on his behalf, whether it is to add particular chemicals to food, the type of housing in a redevelopment area, or whether to invest in public or private transport, until they are too late to reverse.

A third major tendency in the technocracy is the stifling of those changes and innovations which conflict with the single-minded vision of the future envisaged by government and business planners – that is, a larger and glossier version of the present. In this future we are promised that the deficiencies of life which we face now – housing shortages, overcrowded schools, poverty, traffic jams, unemployment and so forth – are to be removed by expanding what we have now. In this future there will be higher production, more roads and cars, faster aircraft, more schools, more overspill housing, in fact virtually more of everything. But the acceptability of this future depends on two assumptions – firstly, that people's ways of life today should be the model for tomorrow, secondly, that there are no practical alternatives to (for example) the pursuit of economic growth, or to the motor-car as a means of travel, or to schools as a way of learning, or to the nuclear family as the basic social unit. Both these assumptions are likely to turn out to be false because they do not take account of discontinuities of change. But by then we shall be saddled with our urban motorways, single family-unit tower housing, supersonic aircraft and the other schemes being decided upon today.

Faced with this situation the individual's main recourse is to protest against choices which adversely affect him directly, or which he considers to be undesirable, or, what is more likely, to merely shrug his shoulders and accept the inevitable. It is hardly surprising that very many people (including planners and designers themselves) feel that any real choices over what happens to them and to the wider community are being denied to them.

INVOLVING THE PUBLIC IN DECISION MAKING

What is being done or being proposed to improve the situation which I have described?

1) One means of discovering the desires and preferences of the population before making a decision is by means of surveys. Surveys of peoples' future desires and expectations are still comparatively rare. A recent example, reported by Abrams (1971), is a questionnaire study to discover popular views in Holland about the likelihood and desirability of certain social changes over the next ten years. (Interestingly, top of the list of desired events was a reduction in social class differences, and bottom a sharp increase in the Dutch population.) However, indirect participation of this sort leaves the choice of questions, and how they are interpreted, with those with the power to choose. For example, in the Dutch survey, although people were asked whether they thought that the standard of living would rise by 1980, they were not asked whether this was desirable; the questioners had already decided upon that answer. A more fundamental objection is that questions of preference about the future can only deal with situations and technologies with which people are familiar and not with the choice of new possibilities.

Today we cannot expect to obtain any more reliable an estimate of the demand for, say, automated transport by polling than would have been found by a survey in the last century of the demand for a hypothetical horseless carriage.

2) In 1968 a committee was appointed by the British government to report "on the best methods, including publicity, of securing the participation of the public in the making of development plans for their area" (Ministry of Housing and Local Government, 1969). The resultant Skeffington Report recommends grafting onto the existing planning machinery means by which the public is informed through exhibitions, leaflets, etc., of proposed development plans throughout their preparation. The public, the report recommends, should have opportunities of commenting on the plan through public meetings and representatives.

This would be a welcome change from the usual conditions of secrecy in which plans are made and implemented. However, all that the recommendations perpermit the public to do is to make detailed comments on plans, the basic concepts of which are already established. The report states that the committee doubted "the need to allow the public to be involved in the establishment of broad aims or goals that the community wish to see achieved", these being best left to the professionals.

Choosing broad aims is surely the most important thing in which the public should be allowed to participate. The details of where a road or supermarket is located are less important than, say, the fundamental choice between road building or public transport or shopping centres versus corner shops.

Since the publication of the Skeffington Report a few planning authorities have held public enquiries into planning decisions, but often these enquiries have been forced upon the planners by public protest rather than being initiated as a matter of policy. An example is the public enquiry into the controversial Greater London Development Plan, which includes the proposed Motorway Box for London. Often these public enquiries have not turned out to be the co-operative exercises envisaged by the Skeffington Committee, but acrimonious exchanges between objectors and planners.

3) A number of proposals have been made, mainly by those involved with futures research, to permit detailed public discussion of a broad range of issues affecting the future. For example, Robert Jungk has advocated the establishment of 'future-creating workshops' for continuous public debate between planners and laymen. He has said (Jungk, 1969):

> "If we want to create a technology dedicated to goals which may be unprofitable in terms of money and power, but important for the 'quality of life' rather than the 'quantity of goods' at our disposal, then the people should have more opportunity to be consulted about the future technology they want and the future technology they would rather reject.
>
> "How will this democratisation be made to work? I see three main avenues:
> a) a continuous mutual learning process;
> b) the education of sufficient intermediaries and interpreters;
> c) the creation of institutions, where experts and laymen meet and co-operate.
>
> "The learning process will have to be instituted at two levels; a) the interaction between experts and politicians and b) the permanent conversation between experts and the larger public".

Realisation of this goal of a continuous dialogue between laymen and experts implies that adults and children be informed about "scientific, artistic and philosophical work in process, anticipated crises and possible future answers to these challenges", through the mass media and the education system, which would have to be oriented towards learning about the future rather than the past. Just as important would be the training of thousands of interpreters to act as go-betweens when experts, laymen and politicians meet. Jungk is hopeful that this will be possible in a future in which more people work shorter hours due to greater automation.

4) A less ambitious means of allowing the public to learn about and choose between alternative future developments is currently under development under the name of Project PLATO (Umpleby and Briggs, 1970):

> "Improved understanding between experts and the public is the goal of a computer-based 'game' being developed at the University of Illinois. The game involves people as 'explorers' of possible future developments. During the exploration they are presented with information about possible occurrences and then are asked to indicate how they would like to change the probability that each of the developments will occur by the year 2000.
>
> "The present version of the exploration describes the future in terms of 40 developments which may characterise the world in the year 2000, such as the development of drugs capable of altering an individual's intelligence, or the appearance of a credit card economy, or the landing of men on Mars. Each development is assigned a certain probability of occurring by the year 2000.
>
> "The explorer is then allowed to 'invest' in each development positively if he judges it desirable, negatively if he considers it undesirable. The computer then calculates the change in probability of the development under consideration that results from the explorer's investment. In addition the computer calculates the secondary effects on other developments that result from the increased or decreased likelihood of this development."

As with the recommendations of the Skeffington Committee this sort of public involvement in decision-making would be welcome. However, once again the alternatives are those selected by the experts, while the public are expected to make choices on the abstract basis of short descriptions of each development and its possible consequences. Abstract descriptions are unlikely to be sufficient to enable people to judge much else but choices between what they are already familiar with. A description is not sufficiently vivid to enable people to choose between developments outside of their experience.

Lasswell (1959) has commented on the need for vivid presentations of possible future ways of life:

> " . . . the methods by which the future is presented do not foster vivid perceptions. It is well known that a trained imagination is necessary before one can perceive a table of figures, a map or a chart. Our perceptions of current and past events are facilitated by the context provided for by the concreteness of news stories, anecdotes and personal observations. By contrast, the charts, graphs and tables that refer to the future lack support. This is a problem especially for non-specialists, since, if laymen are to grasp the meaning of technical communication, they must rely on equivalencies with common experience."

Lasswell goes on to recommend a 'social planetarium' for popularising futures knowledge in the same way as the planetarium has for astronomical knowledge.

What might go on in a social planetarium and how may vivid perceptions about life in the future be created? In other words, how may experience of living in the future be brought into the present so that people may make realistic choices?

5) One such means is the use of the technique known as gaming simulation. This idea is not new. Waskow (1969), for example, has advocated the setting up of futures gaming centres which would "offer experience in living alternative futures to people who are fed up with the present, but have no feel for a workable or desirable society."

At present, progress on participatory games and simulations has been modest, mostly being confined to test groups of non-lay people. There have, however, been numerous attempts to simulate the urban planning process in which participants, in the roles of residents, business interests, planners, etc., bargain with each other over alternative policies in a simulated economic environment, similar to that used in management games. The choices involved are usually of a conventional kind, for example, the grouping of houses, the location of roads, recreational facilities, etc. Although most planning games are intended as exercises for student planners for learning about the planning process, more ambitious games involving residents and planners in clashes over actual policies have been tried (for example see Keyes, 1969).

Other more future-oriented games have involved participants in the roles of various interest groups – the urban poor, the elderly, youth, government officials, business interests, etc., – who propose and oppose and evaluate alternative policies in the light of possible developments such as a reduction in the working week, computer networks, rising crime, age control and so on (for example see Enzer, 1969).

The main value of the games for the participants appears to be in gaining insight into the decision-making process and in learning about predicted possibilities rather than in forming a proper basis for choosing between alternative futures, which requires more vivid experience than can be obtained by discussion.

6) One means of bringing the future into the present for evaluation is the idea of the test city. For example, Jones (1967) has proposed:

> "... test cities of the future would be devoted to the adaptive exploration of all sorts of new ways of life made possible by new kinds of industrial product. Such a city might, for instance, test the response to the combined services of automatic traffic control, variable road patterns, mobile housing, t.v./telephones, education-at-home, and t.v. monitored self-help medicare. It is important to realise that with ingenious methods of simulation it is possible to test reaction to novelties such as these before instead of after the capital investment in quantity production. The chief difficulties of trying out such experiments are those of organisational inertia. The chief requirement for making the test city idea work is the relaxation of many of the legal, moral, administrative and commercial restraints that are relevant only to existing products. This would of course be dangerous, but it would also be exciting."

The high costs and risks involved with the test city concept has meant that we have yet to see a fully fledged experimental city – perhaps the closest to the idea has been seen at World's Fairs, in the various international communities set up by hippies and, in contrast, in the planned Disneyland 'City of the Future'.

7) A method of combining some of the effectiveness of the test city with the cheapness and simplicity of gaming-simulation may be called systemic simulation. The principle of the method is to obtain the total responses and behavioural reaction of test users to simulations of future products and systems in a realistic environment in order to decide whether or not to proceed with their development. The method has successfully been applied to the pre-testing of such devices and systems as voice recognition machines, intelligent computers and various automated transport modes. (For a full description of this method and its development for testing transport systems see Roy, 1971.) The cheapness and simplicity of the method relies on the fact that the simulations need only be rough initially, the important thing being that the full complexity of the human user is retained, and his or her unconstrained responses will reflect the basic requirements that any acceptable future system will have to fulfil.

8) Perhaps the most radically new approach to involving the public in choosing the future involves a reversal of roles; the public becoming the planners.

The effect of role reversal is well exemplified by an experiment in Pennsylvania in which the responsibility for solving urban ghetto problems was shifted from professional planners to members of the Mantua black community. This initiated a process which started to solve the severe problems of unemployment, poor housing, inadequate educational and welfare services which had previously defied solution by professional planners. The project is described by Ackoff (1970):

> "Our approach was based on a few simple assumptions. First we assumed that inhabitants of black ghettos should be given an opportunity to solve their own problems in their own way, that they will not, and should not accept 'white solutions', because whites have demonstrated no particular competance in solving the blacks' problems."

The reversal of roles meant that all the official side had to do was provide local community groups with office and other facilities, finance, legitimacy for borrowing money, and other services or advice which the community asked for. The result was that the community groups set up local industry, employment services, bank loans facilities, local schools, medical centres and many other services. In effect they had taken responsibility for choosing and creating their own future. The community leaders were not professionals and, most important, did not enjoy the protection from errors that outside planners enjoy.

The main obstacle to the repetition of such experiments in community control is the attitude of governmental and local authorities who are reluctant to relinquish their traditional powers and who have little confidence in the ability of non-professionals. This confidence can never be gained unless authorities and professionals become willing to hand over control, at least on an experimental basis.

GETTING FROM HERE TO THERE

Finally, it is highly unlikely that a transition from the present means of choosing the future to that of participatory democracy and local self-government could occur without great resistance from established institutions. Resistance to change is in the nature of large-scale systems. That is why, with increasing demands from various sections of the public to have a say in choosing their future and in protecting their interests, there has been increasing conflict rather than co-operation between the planners and the planned. This has manifested itself in protest movements of all political shades from middle-class amenity groups resisting a road or an airport to revolutionary groups dedicated to the overthrow of all existing institutions. Indeed the political wave of protest is grappling with the issue of self-determination in many guises, emerging as Black Power, Student Power, Tenant Power, Consumer Power and so on.

The strategy adopted, often without knowing, by the various protest movements has been described by Waskow (1969) as 'creative disorder', that is attempting to create today what is desired for the future by obeying the law and order of a future time – a law and order which is likely to be deemed disobedience in today's terms. Established institutions can either condone this or punish it and by doing so generate support or at least awareness of the issues in others. Good examples of this strategy in action are the squatters movement and the various manifestations of the 'alternative society' – communes, rock festivals, the open use of drugs, and so on.

The important rule is to generate the correct level of 'strain' in society by these actions – not so threatening to the present order that complete rejection occurs, not so harmless that no strain is generated. Often acts of protest do not apparently achieve their stated aims and may thus seem pointless. However, as Schon (1970) has said, to overcome the dynamic conservatism of large-scale organisations requires critical levels of energy which have to be reached to precipitate a change of state, which will then occur rapidly. To a certain extent this can be seen to have happened in recent years with the environmental movement which has built up from relatively small academic beginnings to international concern from politicians, planners and public. The environmental movement has sufficient strength, at least for the time being, to halt the building of an American supersonic passenger aircraft, to bring in a programme in the U.S.A. of strict control of exhaust emissions from cars, and, in Britain, to make the choice of a new inland airport politically unacceptable.

Usually, protests of individuals and small groups are insufficient in energy to break through the thresholds of change. They often do, however, produce the strains in society which if they are of the correct level produce systemic changes in public and official attitudes.

The numerous proposals described above for facilitating public involvement in decision-making (electronic referenda, futures games, etc.), are one manifestation of these strains resulting from the response of certain planners to a changing political climate. Many of the proposals described, even if adopted, are likely to be condemned by radical critics as mere sops only offering choices marginally different from each other or just as ploys by those in power to retain it. The issue of who controls the future is likely to be fought over and over again.

A RATIONALE FOR PARTICIPATION

Peter Stringer

I should like to begin by making a few remarks on the expression which has had such magic as to draw us all to this conference – 'design participation'. It has all the ambiguity of meaning to which we are accustomed in the best of our language; which enables us to love and hate, to write poetry and argue. Both words are ambiguous. *Design* can refer either to *the design,* in the sense of a plan for a product, or to the process of *designing. Participation* can mean having a piece of something in common with others – sharing the cake; or doing something in common with others – playing in a game of football. In the first sense 'design participation' must imply sharing the design as a product, in all likelihood the artefact or arrangement which the design posits. In the second sense it implies lending a hand in the process, being one of a design team. There is also a third, and more fundamental, meaning of 'participation'. It can denote *being* a part, rather than *having* or *doing* a part. In this sense participating means partaking of the essential nature of something; and 'design' can be interpreted in either way, as process or as product.

I am assuming in what I shall say that the subject of design participation is a person, rather than a machine, an organisation, or an idea; and that an opposition is implied between laymen and specialists called 'designers'. These are debatable, but fairly obvious, connotations of the expression at this point of history. For the sake of convenience I shall talk indiscriminantly about designs and plans, designers and planners.

THE MOTIVATION TO PARTICIPATE

For design participation to occur it is not sufficient for designers simply to think that it is a good idea. Nor is it a necessary part of their activity, however desirable it may appear. The expression suggests in other fairly familiar connotations a motivation on the part of the general public. As often happens in such matters the desire for public participation has been anticipated before the public has become fully conscious of it. This is often a good tactic, since it gives one a chance to pre-empt their expression of their need and re-interpret it into a handier form! That is probably what I shall find myself doing. But I must attempt to interpret the motivation, since I believe the aetiology of any motivation to participate must be understood if procedures or institutions are to be devised to satisfy it. And some kind of political, social or philosophical rationale is needed for whatever one offers as design participation.

I would point to two major reasons for the motivation. First, a growing recognition that doing is more important than having. Secondly, the ever-increasing rate of change in our surroundings and way of life. The two are interrelated.

The economic goal of obsolescence and the social goal of mobility lays emphasis upon using an object or situation for a restricted period of time, the end of which one can see or anticipate. Because most objects are impermanent and function adequately for a predictably short period of one's life, and because it is actually difficult now to continue doing the same things day-by-day for more than a few years, even if one tries very hard, *change* becomes of paramount interest – and change is process not product, doing or being done to rather than having. Both situations and objects are now pregnant with the possibility of their own succession. For this reason objects lose one of their main characteristics as objects – their stability. In fact critical distinctions between objects and living organisms are becoming blurred. Objects are taking on capacities of growth, reproduction and death. The processes of development, imitation and decay become more interesting than the products themselves. Complaints are also raised that living organisms – and especially people – are treated as objects. Ironically spare-part surgery is introduced at a time when the repair of objects is becoming outmoded.

The most significant thing about the increased rate of change in the objects, activities and ideas which people experience in their own lifetime is not the increase in change itself, so much as the *agent* of change. Whatever relatively small changes occurred in the smoother pre-technological life seem either to have been initiated by the individual or to have been suffered in direct confrontation with another. Major changes were extremely rare for an individual; they were usually initiated by a supreme authority or force, or by acts of

God. Today a large number of both small and large changes in one's mode of living and surroundings *are* effected by oneself. But many others are effected by people with whom one has no direct contact. In the latter case the disturbing sense of alienation is heightened by the realisation that nominally or indirectly one has responsibility for the authority or operation of those others, and that even small changes, in ways too complex to follow, may have far-reaching repercussions for oneself. The economic power that one has at the level of final consumption, and the moral authority which one can exercise in the absence of overriding social or religious dogma or the ultimate legal sanction – and this now includes making much freer decisions about questions of birth, marriage and death – also make it irksome to see an equal power to change being exercised over oneself by others. Both small and large changes in one's life, manipulated from without and with no direct confrontation, become a source of irritation.

MAN'S VIEW OF THE WORLD

There are three principal aspects to this account of why people might want participation. Firstly they have come increasingly to realise their capacities to manipulate their own lives and environment, and to resent the irrelevant manipulations of those whose only authority is one conferred by people themselves. Secondly, in being constantly affected by change they are turning their attention from trying to stabilise the past in the present to predicting and anticipating the future. Thirdly, their manipulations, resentments and predictions are individual. They have their personal view of the world as *they* view it, and it is this which is affected by plans and designs, whomsoever's they may be. The view should be taken to be personal, since there is nothing that guarantees what an individual's view will be – no identity of race, sex, education, age or social class.

These three aspects have been stressed because they are key-stones to a set of philosophical axioms which I believe to be of great value in trying to understand human affairs. I have tried to order my own perceptions through them. I have used them (Stringer, 1970), for example, as a basis for discussing the nature of being an architect. The set of axioms constitutes the basis of the late George Kelly's (1955) Personal Construct Theory. He saw man as essentially active, individual and forward-looking. This is not to say that he cannot be passive, norm-ridden, and retrospective; it is an axiomatic view of his essential rather than his necessary nature. But because Kelly performed the role of a clinical psychologist he tended to see this as a condition which ideally should be actualised as fully and frequently as possible.

He held that a man's view of the world is organised in terms of a system of constructs that are personal to him. The personal construct system enables one to make sense of events around one and order them in relation to one another. It evolves towards an ever more convenient state for enabling one to make more useful and more interesting predictions of future events. A construct system is of course also used to order past events, and it can only be validated by comparing predictions with actual events as they pass. But because of his clinical and therapeutic work Kelly was primarily interested in the evolution of construct systems, and in their capacity to adapt, either in response to changing situations or to produce a different perception of some part of one's world. He believed that a rigid adherence to the validation of a stable construct system and a determination to view the world in a way that led to unvarying and apparently veridical predictions was uninteresting and ultimately maladaptive and unhelpful. This is as true in, say, the physical sciences as in one's personal relationships with others.

DESIGN AS CONSTRUCT EVOLUTION

An evolving construct system, responding to an internal or external requirement of change, often proceeds by propositions typically in the form 'what if' or 'let me look at it as if'. These are a heuristic device for asking about the implications of construing an event in a particular way. These propositions may be shots in the dark or be derived from higher-order propositions in the way in which a classical hypothesis is derived from a theory. Viewed in this way a design or plan can be treated as an indication of an evolving construct system. The hovercraft might be an example of a shot in the dark, 'what if' proposition. It would have been extremely difficult to predict the consequences of viewing transportation in such a way. On the other hand the Boeing 747 or the Concorde more clearly represent *hypotheses* about future travel patterns derived from a theory, however imperfect, of transportation economics. But all three imply not only a change in the way in which one construes transportation; they also imply changes in connected parts of one's construct system – in parts for example, concerned with construing activities sub-served by transportation. Any design or plan which is not simply a straight repetition of an existing one is a new way of viewing a part of the world.

DESIGN PARTICIPATION IN CONSTRUCT TERMS

Design participation can now be looked at again in the various senses that I proposed at the start. In the sense of *sharing* something with others which has been designed, it involves the individual in accepting the imposition on his way of looking at the world of part of another person's construct system. The imposition is not necessarily undesirable. That depends on how welcome it is, and on whether it causes the individual undue strain in trying to incorporate it into his own system or to adjust his own system to accommodate it. The disadvantage is that it is a one-way traffic, and it is difficult for the designer to anticipate the implications of his design – the manifestation of a part of his construct system – for the possibly quite different and numerous systems of others.

Design participation, in the different sense of actively *taking part* in the process of designing, involves the individual either in trying to fit his construct system to that of a specialist, the designer, or in imposing his system on the designer and denying the designer's right or need to have a specialised set of constructs. The latter position is possible but looks unhelpful. The former is back-to-front. If the designer has a specialised and sophisticated construct system, the layman cannot possibly incorporate it into his own without first construing the world like a designer. But he is not a designer, in the specialised sense at least. The designer should rather be fitting his system to that of the layman. But the difficulty about that is that this might prove inhibiting. It might prevent the designer from aiming at radical innovations in construing which are incompatible with the lay systems.

The more fundamental sense of the expression 'design participation' would entail *being a part* of a design or of the process of designing. For people to be a part of the nature of a design presumably means that they are being designed. And this is probably the intention of many designers, who attempt quite explicitly to alter the actions of others through their designed products. Of course, in altering actions they inevitably cause people to reconstrue their worlds. They are tampering with the core of psychological being. On the other hand, for people to be a part of the nature of *designing* is quite a different matter. This recognises not that people should do the designing (I assume here whether rightly or wrongly that they cannot), but that their construct systems are an integral feature of the design process. I assume that the coining of the phrase 'public participation' in itself suggests a denial of the sense of the expression which amounts to people's lives being simply the object of planning. Presumably also there is no intention, at least on the part of the authorities, to have the public deny the planners their role or usurp their function. One is thus left with the sense in which the public are an integral part of the essential nature of planning. Of course, the very fact that the agents of changes brought about by planning are employed by and responsible to representatives of the public should also guarantee that. In laying so much stress on the more fundamental sense of 'participation' I have taken the argument well beyond *having* and *doing* onto the realm of *being.* I should make it clear that while the transition from an interest in having to one in doing is scarcely yet under way for many of the population, the further transition to being is still a matter of primarily philosophical interest.

COMMUNICATION

I have said earlier that a plan or design constitutes part of a specialist construct system. If it is to be accepted and put to use, there must be a congruence between the plan and the user's constructs, unless considerable strain is to result. There are various ways in which this can be achieved. The congruence can be formed at the user's despite by physical necessity or superior authority; he can be placed in a position where he must reconstrue events if he is to maintain anything like his preferred way of life. This is often called 'adaptation'. People may come to reconstrue a tower dwelling as having all the essential properties of home because they have little chance of doing otherwise, unless they are to suffer hardship and disruptions in other parts of their construct system. Or the congruence can be formed insiduously. The plan can be ascribed properties that are illusory or relatively trivial in order to make it fit the public's view of the world. This is most common in the field of consumer product design.

Neither of these eminently convenient tactics are morally acceptable, except perhaps in rare and exceptional circumstances. A third method of achieving congruence is for the planner to apprise himself of the public's various construct systems, and, treating them as given, to find ways of making his system maximally congruent with theirs. This is akin to what I have been doing in a current research project, which has involved asking a sample of the public to construe a number of alternative plans for redeveloping their local shopping centre. It is quite apparent that they can do this. They produce a relatively large number of constructs and they show a substantial measure of agreement with one another. Their constructs, however, are not those of the planners in many important respects; nor, interestingly, do they match those of self-appointed watchdogs in local amenity societies.

The research project is an idealised and costly means of learning about how people construe possible future environments. To pursue the ideal, though this is *not* part of the project, one would expect the planner to find ways of subsuming their constructs to his own, and thereby to produce a plan which reflected both their sets of views as to what would be a convenient and interesting environment for a shopping centre. This would be very difficult. But it might be a reasonable kind of task to require of a highly-trained and highly-paid professional who has elected to work in the public service. It might be claimed that this ideal is in fact what does happen in planning offices. If so it must be by osmosis, since virtually no visible means for collecting the necessary information exists. And in many notorious cases the planning membrane has obviously not been thin enough for the public's constructs to be transmitted.

EDUCATION

But even this ideal falls far short of what one would hope for in contact between two construct systems. The contact is only one-way. There is no means by which the public can adequately inform the planner of their view-point; they must wait to be asked. And it is very rare for the public to ascertain what the planner's constructs are. They are either not told at all (and are unable to divine them from the plan itself for lack of expertise), or they are told and are unable to understand, the constructs being sophisticated and complex and expressed in unfamiliar language.

The fourth means, then, of achieving congruence between the two viewpoints, requires that there be *full* two-way communication. And because one party has a set of constructs that are more complex, it also requires an expository or educative process in which the complexities are made fully intelligible to the public. When aid is given to an undeveloped country, it is usual to ensure that some of the population understand both the function and the long-term purposes and implications of the new financial and technical resources. In developed countries very few people understand measures that are taken on their behalf and are bought from their labour.

A proper education is not a matter of learning by a particular set of conventions. It is a matter of trying on a variety of points of view to discover which gives the most convenient and interesting anticipation of events. If the viewpoint which is the subject-matter of this education is to become related to the individual's personal construct system, he needs to test it in real situations, to become personally involved with the viewpoint, and committed to its implications. This cannot happen if it is merely expounded in the abstract, in relation to situations in which the learner plays no role. This adds up to saying that if the planner wishes to achieve congruence between his terms of reference and the public outlook, and if the public wish to understand and be understood in planning affairs, a context must be found in which the public, as individuals, can be committedly involved in acting for the future in a way that could make such institutions relevant.

While questionnaires, representative consumer panels or referenda are quite inadequate to give one a satisfying sense of involvement and commitment and to allow the individual to develop a more complex and highly evolved personal viewpoint on the world, I am not suggesting that one should go to the extreme of having the public usurp the planner's present function. The layman is very experienced, and often quite good, at planning other parts of his life. What is necessary is that he should be able to exercise that talent at some level of the more technical planning of his environment. It seems to me that this will only be possible with a radical redefinition of what we understand now by designing and planning.

ACKNOWLEDGEMENT

The preparation of this paper was supported in part by a grant from the Centre for Environmental Studies.

PARTICIPATION IN PLANNING DECISIONS

Peter Levin

A discussion of the question of participation in planning decisions ought, I think, to begin with a definition of terms. I should like, however, to defer defining 'participation' until later, and at this point simply to explain what I mean by 'planning' and 'decisions'. The term *planning* I take to denote the act of "arranging beforehand", as the Concise Oxford Dictionary has it, and I shall be concerned here with 'physical planning' – the arranging beforehand of changes in the physical environment, as exemplified by the building of a new town or motorway, or the carrying out of a comprehensive redevelopment scheme. Before such an action is embarked upon by central or local government, or by some statutory body (and I shall confine myself here to planning by government), there always takes place what I can only describe as acts of forming a resolve. In 1965 the Government resolved to build a new town in the Leyland-Chorley area. In 1970 the Minister of Housing and Local Government resolved to make a designation order for the new town. Between these two points in the planning process lay many other acts of forming a resolve. Sequences of this kind characterize all governmental planning processes.

Acts of forming a resolve conform to the definition which I wish to adopt here for the term *decision:* a decision is an act of choice which generates commitment to a specified course of action. The course of action that is chosen thus possesses two important properties – a) it possesses specificity, and b) commitment is attached to it. It has to possess specificity in order that it may be distinguished from other possible courses of action: the higher its specificity the more closely it will approximate to a single blueprint for action. The commitment attached to the chosen course of action is a measure of the decision makers' resolve that it rather than some other course will be implemented. The strength of commitment is itself to be measured in terms of the penalty perceived by the decision maker (or makers) to be associated with rescinding a decision and making a different choice. Commitment thus constitutes a perceived incentive to persist. Two things are to be noted. First, commitment is a relative quantity: the penalty associated with changing from course A to course B may be different from that associated with changing from A to C. Second, the existence of commitment to a course of action does not imply that the choice is irrevocable: a decision maker will sometimes consider it worth paying the penalty involved.

The point at which one of the parties first resolves to try to bring about a change in the physical environment and the point at which implementation of the change is complete (or the enterprise abandoned) mark the beginning and end of the planning, or decision-making, process, and between them lies a whole host of decisions – about making investigations, earmarking finance, informing those who might be affected – each of which carries commitment to the action it specifies. Thus the decision to make a particular investigation carries commitment to the actual making of the investigation. But in addition each such decision and action may play some part in increasing the specificity of the development proposal and in gradually building up commitment to it, as I shall show later.

THE PLANNING PROCESS

If we are to discover opportunities for people to influence the specification of a planning proposal, we must look for them in the planning process. What we need first of all is a picture or model of that process, hopefully one which reveals discontinuities and differences of some kind, for if all parts of it are alike it will be hard to infer that one part of it affords better opportunities for the exercise of influence than does another. There is of course one particular model that has been around for a very long time, namely the 'formal authorization equals decision' model. According to this, before the formal authorization, nothing has been decided: after it, nothing can be changed. There can be few residents' associations, preservation societies or protest groups of other kinds that have not at some time or other encountered this view of decision making from a local authority or government department. On the 'formal authorization equals decision' model, commitment leaps at a stroke from zero to 100%. It is a manifestly inaccurate representation of what really goes on.

I wish to use here an alternative model, a little less elementary and – I hope – more realistic. Its basic feature is that it resolves the decision-making process into three component parts. These correspond recognizably, but not perfectly, with the 'administrative', 'technical' and 'political' processes of not unfamiliar usage: the correspondence is close enough for it to be reasonable to use these labels. These three processes, which are set against the background of an 'event stream' (events which may change the perceived need and scope for action), are made up of sets of acts and activities which are distinguished by their primary purposes. But they have in common that they contribute to the build-up, over a period of time, of commitment – commitment to an increasingly specific course of ultimate action. Thus, I am postulating that commitment and specificity progressively increase, and do so by discrete increments, rather than there suddenly appearing a massive commitment to a highly specific course of ultimate action.

The *administrative process* comprises those acts which must necessarily – by virtue of logic or of prescribed rules – be performed before a particular change in the physical environment can be made by a particular agency.

There are three main ways in which the administrative process generates commitment. First, many administrative acts require a visible staking of judgment – visible, that is, to the associates and colleagues of those who make the judgment if not to the public at large. An administrator may have fought his hardest, and staked his judgment, to convince his colleagues of the urgent need to earmark money for a particular project. To reverse a decision taken on such a basis may therefore involve for him a considerable loss of face, or loss of credibility – a penalty in fact. Commitment has been generated. Second, an administrative decision will often provide the basis for the expenditure of scarce administrative and technical resources – the time and effort of highly educated people on whom the success of the project may depend but who have the ability to refrain from giving their full energies to a task. If such decisions are subsequently changed, not only may much trouble and care be wasted but loyalty and keenness may be forfeited. The prospect of paying this penalty will obviously deter changes; the decision makers will have incurred commitment. And third, the delay occasioned by such a change is likely to cause a delay in the construction of the new town, say, with a consequent penalty (although perhaps not without eventual compensation) to be borne by those who remain longer in poor and over-crowded housing in the big cities. Those who reverse an administrative decision may, depending on their reason, bear a burden of blame for inflicting this penalty. The prospect of this will act as an incentive to persist. We see, then, that in the course of the administrative process some of those involved in it necessarily incur commitment to a certain ultimate course or range of courses of action, and it is true also to say that in an uninterrupted administrative process commitment will progressively increase as the process goes on. Indeed, by the time that the proposal is ready for a formal authorization, a great deal of commitment may already have been incurred.

The *technical process* consists of acts and activities of acquiring, generating or processing information that relates to the perceived need or scope for action to change the physical environment. Just as the administrative process enables an action to be taken, so the technical process enables that action to be specified. Initially those involved will have some idea, perhaps scarcely quantified, of certain problems which seem to require action to solve them; they will probably have an equally rough idea of what kinds and scales of action there is scope for. In the course of the technical process 'desired outcomes' or goals will become more specifically defined and limitations on action more fully explored and appreciated. Relationships between action and outcome will be identified, implicitly: they will be used to discover what the outcomes of a given action will be (these then being compared with the desired outcomes) and to discover what actions need to be taken to achieve a given outcome (these 'desired' actions then being assessed against the identified limitations on action). Characteristically, as the process goes on, the initial picture of the action context – a subjective picture in the mind of each participant – develops, becoming more ramified, as more variables are added to it and the limits upon the values that they may take are more precisely established. At the same time, the picture becomes more consistent, as certain possible courses of action are rejected because they are perceived as not enabling goals to be achieved, or certain goals are abandoned because they call for a course of action that lies outside the permitted limits, or certain constraints on action are relaxed because they would make a goal unattainable. Thus preferred and highly specific courses of action emerge. It may be necessary to establish relative preferences, which will be done by applying a common criterion (measure of outcomes) to each course of action on the 'short-list'. Cost-benefit analysis is a particularly explicit way of doing this.

Commitment is generated in the course of a technical process just as it is in the course of an administrative one. However, while the effect of administrative decisions and activities is generally to ratify, and attach commitment to, the specification that is the output of the technical process, the latter itself operates to raise the specificity of a proposal as well as being a generator of commitment.

The analogue of the confrontation between administrators, in which a staking of judgment may be necessitated, is the confrontation that is liable to occur within the planning team, especially when it is composed of representatives of different disciplines, or firms, or authorities (e.g. local councils). If the various representatives have different ends, so that an element of conflict is present (and the process becomes a technical/political one), the confrontations that take place may be very formal. Often only a limited degree of responsibility will be delegated to the representatives, so that they frequently have to refer back to their principals: communication lines are extended and it is exceedingly difficult for the separate parties to come together to explore each other's minds and discuss questions at a tentative level. When meetings occur they are more formal, and may have more rigid agendas, than they otherwise would be. Great importance is invested in the minutes of meetings, and those involved find themselves pressed to make firmer statements than they might wish and subsequently being held to them. They are under pressure, in other words, to commit themselves to the results of only limited exploration. The resulting proposals are likely to be more highly specific – probably in terms both of range and detail – than they would otherwise be, let alone different.

In the technical process too there are procedural decisions, which provide the basis for the expenditure of scarce man-hours by valued and not-to-be-offended technical staff. The same deterrents to reversing these decisions will be faced as in the case of the administrative process. Many of these procedural decisions will have been imposed by the need to cope with a complex mass of data rather than administrative necessity. Decisions as to the methodology to be adopted, or as to assumptions which are to be made, are examples of this. Such decisions are likely to have the effect of narrowing down the field of exploration: hence their effect is to increase specificity as well as commitment.

It is also invariably the case that the town planner, like most other people who have to handle and reconcile a mass of information in a problem-solving situation, very easily acquires a personal commitment to the solution he personally prefers, and may begin to do so quite early in the process (indeed he may approach it at the outset with a firmly-held preference). The growth of a personal attachment to a preferred solution is perhaps inevitable, given human nature. It would certainly seem to be fostered by the fact of life that factual information on its own cannot determine what solution should be adopted (an 'is' can never be transmuted into an ought'). The gap left by the information in this respect is often filled by the prevailing professional ethos, which reinforces instinct by placing a premium on novelty, on producing a solution which can readily be distinguished from those of other people and is identifiable as the planner's own work. Pressure to produce a novel solution may mean that very little effort is put into exploring alternatives, and that an approach developed by another planner may be ignored even though it may be applicable in the context. And if one or two solutions initially seem very promising, the planner will naturally choose to investigate them in detail, something that limitations on resources will not allow him to do with every potential course that appears to be open. If his most promising solution is consistent with his picture of the need and scope for action, this will naturally give cause for satisfaction. Now, if at this point, a new piece of information, conflicting with this picture, comes to his notice, one of two things will tend to happen. Either he will be thrown into confusion (of which rash behaviour will be a symptom) or, if he can, he will dismiss the new information. Both types of reaction are known: both are plausibly explained by the existence of a strong psychological commitment by the time the new information comes to hand. Once again the growth of commitment will be associated with an increase in the specificity of a proposal.

Finally, commitment is generated in the course of a technical process, as it is during an administrative process, by virtue of the penalty – in terms of the additional technical work needed, and the delay suffered by the ultimate beneficiaries – attendant on rejecting the proposal that emerges. In such a situation, especially where the planning process has already extended over a long period, the over-riding wish of many of those involved may be simply to end the uncertainty, and whatever plan is available may be seized upon purely because it fulfils this purpose, and despite its containing inadequacies which further study might have resolved.

An individual or group of people engaged in making administrative or technical decisions or carrying out administrative or technical activities essentially does so on his or its own, although they have been influenced by interactions with other people. It is the political process in which interaction takes place. The *political process* may be defined as consisting of acts and activities directed by one group towards influencing the views of one or more others, generally in the direction of strengthening or weakening support or opposition for a proposal among them. As with the other two processes, commitment is generated in the course of it.

The focus of these acts and activities is invariably the specification of the proposal, but to bring about a change in it, it is usually necessary to lower or counteract the existing commitment on the part of the planning agency and to cause it to incur commitment to the change. Thus a group opposing a proposal may seek to convince its proponents that the proposal is based on incorrect 'facts', or that a particular criterion has not been given the weight it deserves, for example the suffering that would be experienced by a particular section of the community. If the attempt succeeds it will be by giving the proponents an incentive to make changes, an incentive that will outweigh the existing commitment. Instead commitment will be incurred to the amended proposal. Where the opposing group is in a position to exert an obligation on the proponents the political activity may take the form of bargaining over the specification. If a bargain is reached the effect will be that the proposal is changed (and may be more specific as well if safeguards and provisions are incorporated) and that the proponents are more highly committed than they were previously, since to abandon the amended proposal involves breaking a bargain with these opponents, which making the change did not. Finally, engaging in a political process may strengthen the proponents' commitment to the existing proposal. Vigorous insistence in public that there can be no going back obviously has this effect.

MAXIMISING THE SCOPE FOR PARTICIPATION

Now, what is the relevance of this analysis to the question of participation? If we equate participation in a process with 'having a share in it' (to use the Concise Oxford Dictionary's definition), then it follows that in the political process alone – as the only one of the three that involves interaction between different groups – is participation possible. If we wish to design a decision-making process that will allow for maximum participation, this has a two-fold implication. First, it implies that the political process must be designed to incorporate maximum participation. Second, bearing in mind that each of the three processes may have the effect of building up commitment to a specific proposal, it implies that the administrative and technical processes must be designed in such a way that they do not wholly determine to which specification commitment is generated nor the total amount of commitment generated, but leave the maximum scope in these respects to the political process.

Let us discuss these implications for each process in turn.

The administrative process

I referred earlier to three ways in which the administrative process can act as a generator of commitment – through the visible staking of judgment; by necessitating the basing of expenditure of valued administrative effort on a decision and thereby effectively 'freezing' it; and by involving the deferment of benefits if a decision is re-opened. So far as the staking of judgment is concerned, it would seem to be a defect of an administrative procedure that it should put, as some do, an onus upon an administrator to become strongly and personally committed to a project before its feasibility and its political repercussions have been fully explored. It goes without saying that because of the competition for Exchequer funds they have to be earmarked well in advance. But what earmarking there is cannot absolutely guarantee that finance will be available: we all know that in times of economic crisis no environmental project is safe. One would suggest, therefore, that earmarking at an early stage of a proposal's investigation should be no more than tentative, that the burden of justification thrown on administrators should be no more than is appropriate to the state of knowledge and public discussion. 'Firming-up' can perfectly well follow at a later stage. This principle is already followed with schemes for motorways and trunk roads: these schemes are put first into a 'preparation pool' for detailed investigation and evaluation and then transferred into the 'firm programme' in accordance with their relative priorities. There seems to be no reason why the maturity of a scheme for a new town, for example, should not be similarly recognised.

Given that administrative and technical resources are scarce, which necessitates decisions as to how they are to be deployed, and that they are 'locked up' in human beings of generally well-developed intellect on whose loyalty and devotion to duty the proponents depend and who are therefore not to be alienated by having the results of their efforts wasted, there would seem to be no way in which the freezing of a decision by virtue of the work (and further decisions) subsequently based on it can be avoided. If this is so, then for participation to take place it is necessary that the political process be brought into play before that decision is taken. To take an example, it is obviously necessary that when consultants are commissioned to prepare designation proposals for a new town, they must be given terms of reference. These will inevitably commit the consultants to producing a proposal that conforms to them, and in the normal course of events the Minister will find himself committed to a similarly conforming designation order. If the limitation imposed by the terms of reference is to be challenged, it must be done before (or at worst very soon after) the decision as to terms of reference is made. This, then, is the time for the political process to be operating.

The third way in which the administrative process can act as a generator of commitment is by virtue of the fact that reopening a decision may involve repeating various administrative steps with a consequent delay in the attainment of the ultimate benefits, blame for the delay being perceived as a penalty by those who would be responsible for reopening the decision. Here, again, the argument must be that if participation is to be a reality the political process must be allowed to operate before 'time-sterilizing' decisions are taken. But we should note that the penalty – and thus the commitment – could be lessened if there were alternative uses to which the resources that would otherwise lie unused could be put. Thus if the delayed project were one of a 'preparation pool' and it were possible to divert the resources to another project, the benefits from this could to some extent be set against the benefits lost by the delay, although it must be true to say that only very rarely will those who lose by the deferment of one project be identical with those who gain from the bringing forward of another.

The technical process

In the course of a technical process commitment is generated and the specificity of a proposal raised. Three kinds of act and activity were identified that had this dual effect – the visible staking of judgment; the expenditure of scarce human resource following a decision; and the growth of a psychological attachment to a preferred specification. And again – as in the case of the administrative process – commitment may be generated by virtue of the fact that the deferment of benefits is involved if a decision is re-opened.

If the growth of commitment arising from confrontations within the planning team is to be minimized, it would seem that the formality of the confrontations must also be minimized. This has implications both for the approach to the problem and for 'management style'. For an initial period it would seem to be wise to put the emphasis on exploring the problem rather than on producing the solution: it tends to be a less divisive activity, and allows the team a 'running-in' period, which is essential if a 'team spirit' is to develop. Such a spirit both contributes to the lessening of formality and provides a cohesive bond that will help to resist the fragmentation that is liable to occur when conflicts over solutions arise. It also enables a freer management style, one which can – for example – allow a good deal of freedom for junior members to question what is done: in a cohesive situation such questioning is least likely to be intended, or interpreted, as a challenge to the leaders of the team. Accordingly a wider exploration is likely, unhampered by premature growth of commitment, and the outcome may be the emergence of a range (or a wider range) of alternatives. Thus the outcome would be of lower specificity than it might otherwise.

The procedural decisions taken in the course of the technical process provide, like those taken in the course of the administrative process, the basis for the expenditure of scarce human resources. Once again, there would seem to be no way in which to prevent the work done subsequently to a decision from 'freezing' it: hence we are forced again to the conclusion that crucial decisions need to be taken in the context of the political process if the scope for participation is not to be fettered by them.

If the growth of personal attachment to a preferred solution is to be held in check, ways must be found of relieving or counter-balancing the pressures on the planner to produce a novel or distinctive solution. Relieving the pressure would seem to be the more difficult thing to do. Although a much greater emphasis is placed nowadays on methodology than it used to be – not least because planners are having increasingly to defend their proposals against alternatives – there is also a greater emphasis on inventing a novel methodology. We have moved only from monumental planning to monumental planning reports. For a counter-balance to the pressure of the professional ethos one must obviously look to the political process. I shall come to this later, merely reiterating here that if several alternatives are under genuine consideration on such an occasion far less commitment is called for on the planners' part than if there is only a single one, in which case not only are they likely to be already committed but they will almost certainly find themselves in a defensive position, which will force them to increase their commitment still further.

If to re-open a decision already made necessitates a further instalment of the technical process, and this will involve the deferment of ultimate benefits, then there may exist strong commitment to that decision. But the amount of further study that is necessary can be cut down if some (or indeed all) of it has already been made. This obviously involves carrying a number of alternatives forward to the same stage of detailed investigation, which will naturally take more man-hours than if it is done for only one, but this may anyway be necessary for the purposes of evaluation. Such a procedure was in fact followed by the Roskill Commission, which investigated four alternatives in great detail. The Government's rejection of Roskill's preference was certainly facilitated by the fact that the information necessary for comparing Foulness with Cublington was available, and a further study was not necessary.

The political process

It was the political process alone that was identified as offering scope for participation: we can go on to discuss how its potential might be realized. We should bear in mind that there almost always is a political process associated with the making of planning decisions, albeit one often restricted for much of its duration to a small number of bodies. So if there is to be wider participation, the question is one of how to enlarge the number admitted to it, rather than to create such a process from scratch. The first prerequisite is to make information widely available, both the information that has been obtained in the course of the technical process relating to the need and scope for action and information about the levels of commitment that have already been reached. The next step is to provide for interaction. The context for the political process is set by the current specification of a proposal and the already-existing level of commitment to it, and offering people the opportunity of participation implies allowing them the opportunity of influencing the way in which these two quantities subsequently develop.

Participation in influencing a specification must take the form of contributing and exchanging factual information and subjective reactions, which will be based on personal values. We have so little experience of participation experiments that it would be foolish to try to set out a sure-fire recipe for participation in influencing a specification, but from the experiments that have been tried it has become apparent that certain constraints are commonly imposed – sometimes deliberately, sometimes not – to its severe detriment. In default of a sure-fire recipe, it may be useful to warn against these pitfalls, which might be labelled over-abstraction, undue limitation of the field, and inadequate vehicle.

Over-abstraction occurs through the insistence of planners in particular on advancing proposals at the level of general strategies – which is good, but unfortunately they tend to decline to discuss the real-life implications of those proposals. This tendency has been very strongly reinforced by the introduction of 'structure plans', which have no ordnance survey base, as components of statutory development plans. The public's reaction is invariably to ask "how does this affect me?", thereby earning the disapproval and contempt of the planners. Personally, as I have said elsewhere (Levin, 1971), I would regard this reaction as a perfectly understandable and legitimate attempt to discover the reality that lies behind the abstraction. It possibly also reflects a fear that an abstract structure plan, say, may lead irrevocably to an all-too-definite local plan – in other words, that a hidden commitment to a high-specificity local plan is being built in. This fear may sometimes prove ultimately to be justified. It would in any case seem sensible to explore the concrete implications of abstract plans before commitment becomes attached to them, and to do so would certainly make participation easier.

Limitation of the field usually tends to be imposed on a participation process if certain decisions have already been taken and it is not within the competence of the authority sponsoring the participation to reopen them. The objective of the process is accordingly limited to discussing the matter in hand. But sometimes these limitations may be more arbitrary – or undue – as when the authority insists that it will not tolerate 'destructive criticism' of its proposal.

Inadequate vehicles for participation are legion. They include the one-off public meeting, inadequate because the dialogue is forced into a very formal mould and because participation needs to be an on-going process as commitment and specificity grow with the formulation, development and examination of new solutions; the "send us your observations and we will tell you if and why they have not been accepted" formula which suffers from the same drawbacks as the one-off public meeting; the questionnaire, which tends to reflect the conceptual framework and attitudes of those who draw it up rather than those who fill it in, and is again essentially a one-off rather than an on-going vehicle; and public inquiries into objections to draft designation orders for new towns, at which the proposals are almost never defended by those who prepare them and the arguments and alternatives put forward by objectors are never tested in cross-examination. As regards the first three of these, David Donnison's (1970) suggestion of a standing randomly selected community panel which would choose questions for surveys and public discussion would seem to be an improvement. And even the conventional development plan inquiry format would be an improvement on that of the new town inquiry.

REACHING A COMPROMISE

Let us assume that the widest range of possible specifications for the proposal is exhaustively discussed and examined, with the fullest participation of interest groups and the public. What influence will the contribution of these participants have? Now, the exertion of influence will be manifested in terms of commitment as well as specification. If they accept – or, more positively, actively support – the proposal that the planners prefer, then the process will generate commitment to that proposal. But what if there is conflict? The participants will need to persuade the planners to share their values, and for this the submission of information is a very weak tool. Only rarely, as when those who fill in a questionnaire declare themselves unanimously against being re-housed in tower blocks, can information achieve this end. What a unanimous consumer response does is to place the planning agency under an obligation, an obligation to take account of the views expressed, and it would seem that only be exerting an obligation can, for example, a commitment to a novel solution be outweighed. The ability to exert an obligation – in other words, political power – without having to deliver a quid pro quo is generally accorded only to representative bodies. This is why, in my opinion, we need neighbourhood councils or similar bodies if the inhabitants of our towns and cities are to be able genuinely to participate in the making of planning decisions.

Given that the complete range of interests is represented in the political arena, and that many incompatible demands and proposals will be put forward, the task of the planning agency will be to explore the possibility of trade-offs and compromises. It will be helped in this by the fact that conflicts and areas of disagreement will have been made explicit. And where no compromise is possible, and the agency decides that one group has to give way, it will at least be clear to everyone why – and clear too if one particular interest group consistently comes out on top or if one is consistently over-ridden. Decision making through a process of this kind should help to ensure that compromises and trade-offs are fully explored, with most people getting a fairer deal than they would without it. The negotiation of a compromise or trade-off between parties will generate a commitment to the specification arrived at: it is, I suggest, the optimal way of generating commitment during a planning process. Of all those ways of generating commitment that I have mentioned, it conforms most closely to the rules of 'natural justice'.

EFFECTIVE CHALLENGES

So much for conditions that favour participation. What are people to do when a planning agency puts forward a proposal that would affect them adversely, and does so in conditions quite unfavourable to participation, that is to say by putting forward a highly specific proposal to which a great deal of commitment has already been generated in the course of the administrative and technical processes? Clearly they need to challenge both the specification and the commitment. With regard to the specification, it is necessary to develop their own self-consistent picture of the need and scope for action, and the chance of success will probably be enhanced if it has a certain amount in common with that of the agency's planners and if, for example, the need for some action can be accepted without invalidating the challenge. It is worth while making use of the expertise of professional planners, especially as this will give the challenge authority and the appearance of that mythical quality, objectivity.

A well-mounted challenge, if it is directed to genuine weaknesses in reasoning and technique, may make the agency's planners wish that they had never thought of the proposal. However, in a public confrontation they may be unwilling to admit to being in error, and in defending their proposal – adducing perhaps their 'professional judgment' in support of it – they may become further committed to it. The challengers' aim must be to prevent this additional commitment – and indeed that already incurred by the planners – from extending to their political masters. To do this it is necessary to weaken the latter's obligation to the planners, an obligation – imposed by contractual and organizational bonds – to heed their professional judgment. In part this can be achieved by demonstrating any weaknesses in the proposal and emphasising that they should have been foreseen; and if the planners can be shown to have been keeping the politicians in the dark over certain matters, for example, this too will lessen the politicians' commitment.

The next step is to provide the politicians with a positive incentive to adopt an alternative proposal. This aim will be furthered by organizing a group representing as many as possible of those affected, or a coalition of groups, sufficiently representative for the politicians to feel under an obligation to heed its views, and to perceive some value in conciliating it. It is now time for a respected elder statesman to suggest to the politicians that they themselves would be showing statesmanlike qualities by recognizing the strength of public opinion and, while not for one moment accepting that there are deficiencies in their planners' work, in acknowledgment of it to adopt a compromise if appropriate or to give the proposal further thought.

CONCLUSION

In this paper, I have tried to identify ways of creating opportunities for participation in planning decisions. I have frequently been drawn back to the same point: the greater the level of commitment among administrators, planners and politicians, the greater the inhibition to participation. I have not rehearsed here the arguments why participation is desirable. I have taken its desirability for granted and hence much of my argument has been devoted to ways of minimizing the generation of commitment in the course of administrative and technical processes. I believe, however, that there has to be a certain level of commitment to a proposal in advance of any widespread participation if the proposal is to have any chance of getting off the ground. It is a fact of political and official life that sensitive negotiations cannot be conducted in a glare of publicity, and a fact of human nature that few people are willing to expose themselves by making a public initiative until they have assured themselves of at least some support: commitment is inescapable. We have reached a new formulation of the old dilemma, how to balance administrative efficiency against natural justice. In giving the scales a push towards natural justice, we must not forget that a balance must be kept.

GAMING – SIMULATION: AN APPROACH TO USER PARTICIPATION IN DESIGN

Ignacio Armillas

Traditional methods of design are proving inadequate to the task of dealing effectively with the increasingly complex nature of the man-made environment. In a search for better methods the barriers that separate specialists in the various design fields are crumbling. Now the last barrier – that between designer and user – is beginning to fall. Design is at last losing its mystique of lofty artistic ideals which have enveloped it for so long.

Planning was the first design field to recognize the need for the participation in design activity of those who, for better or for worse, must live with it and perhaps suffer the results. It also pioneered attempts to bridge the gap between planner and 'planee' in a systematic way. One contemporary result of these efforts is 'advocacy planning'. Although architecture has been concerned with user needs for a long time, this concern has seldom been translated into active participation of the users in the design process. Industrial design has also been very much concerned with the needs and desires of the user. These users, like the users of architecture, rarely participate in the design process in an active manner. Decision-making has been and largely remains the sole domain of the professional designer.

Unfortunately, we lack a methodology for incorporating the user into the decision-making process. If user participation in design is to be a meaningful and productive activity then ways of involving the user must be developed. This paper is about one such way: the use of gaming techniques. URBANISTA, a gaming exercise being developed at the School of Architecture of Nova Scotia Technical College, is a particular example of the application of these techniques.

Two terms relevant to this discussion are games and simulations. Games are formalized conflicts where the conflict is between two or more parties which at any specified time have a range of options concerning their actions relative to a given situation. The parameters on the range of options are prescribed by the rules which are assumed to be known by all players. Play is basically a sequential decision-making exercise structured around a model in which the participants assume the role of operating the simulated situation. Games usually rely on simulation models, but simulation models do not rely on gaming.

Simulations are analogues of real life phenomena, a range of abstract representations of real world components. Simulation models allow us to manipulate a series of variables in a synthetic environment and, unlike gaming which always involves players, simulation does not necessarily entail the involvement of individuals.

Only recently have the traditional disciplines of the social sciences recognized the potential of gaming and simulation techniques as aids to research and analysis. Foremost among the early efforts in this area were those of the political scientists who succeeded in developing sophisticated simulations of conflict resolution in international politics. Gaming continued to gain acceptance in other fields as is evident by the number of exercises developed in such varied disciplines as conservation, economics, watershed management, hospital and hotel administration.

Although a recent entry into the ranks of those subject areas that are utilizing – or at least exploring the potential of – gaming and simulation techniques in their own substantive areas, planning games have already reached a high degree of sophistication. Planning games replicate the salient characteristics of urban systems in a synthetic environment in which time and space are greatly compressed. Most planning games permit participation by non-planners and in fact many are used as heuristic tools to give the players greater understanding of the problems confronting contemporary urban societies. CLUG (Feldt, 1968), Metropolis (Duke, 1964), and City (Environmetrics, 1971) are three games which fall into this general.category.

These games have been played by a number of different groups including university students in both informal and classroom situations, professional planners in conferences, planning faculties, developers (including new town developers), city councilmen, ministers at a church conference, and other groups of laymen.

CLUG has even been played as a parlour game in social situations. One model of City was used as an aid to teach a group of Washington, D.C., ghetto-dwellers about the metropolitan area as a whole. Although the players had only an average of a tenth grade education, the game was successful in terms of providing the players with an increased understanding of the urban system.

The games mentioned above are not intended to be design tools. Nevertheless, gaming-simulation techniques have the potential of involving the user in the design process. This potential is derived from some of the characteristics of gaming and simulation models such as: the simplification of a situation or system through simulation; the heuristic qualities of gaming which appear to be effective in teaching dynamics of systems without requiring great knowledge of facts and theory from the players; the compression of time enabling quick feedback and the reduction of costs to a mere fraction of the real world cost.

In addition, there is a great affinity between gaming-simulation and design. Gaming – the exploration of strategies to solve conflicts through the use of simulated situations – is essentially a sequence of trade-offs. Likewise, design – the development of strategies to solve conflicts – is a process of compromises. Though the terminologies used by gamers and designers might be somewhat different, the differences are mainly semantic.

URBANISTA is a gaming exercise that has been developed as a design aid for groups of players who include both designers and users. As a means of exploring applications and consequences of design strategies, the game has been extensively used in an urban design course at the School of Architecture in Nova Scotia. The players in this urban design laboratory included students, staff members and outside guests.

The gaming format of URBANISTA provides the players with an approach to the problem which requires the explicit formulation of objectives, the development of strategies and the systematic evaluation of the consequences resulting from a particular strategy.

URBANISTA is a game that can accept a variety of environmental design problems ranging in scale from an individual building to new towns. So far it has only been applied to one problem: the development of certain land on the urban fringe of Halifax, Canada, that has until recently been reserved as a source of water. In URBANISTA the players can 'build' this area as an extension of the city, the various factions within the game – government, developers, users – acting according to the values and self interests of their roles. Therefore, while the overall objective is to develop the land, each faction influences development in a different manner.

Six teams represent three factions: Capitalists, Populists and Environmentalists. Capitalists seek to maximize economic return on their capital invested; Populists represent socially-oriented action groups concerned primarily with human values; Environmentalists are concerned with the quality of the natural and/or urban environment. The efforts of the six teams are integrated and implemented by a seventh group of players which represents the government. This team always consists of four players, each independent of the others. These factions, of course, are extreme simplifications of objectives in the real world that are not necessarily mutually exclusive. The choice of particular factions with clearly defined characteristics and value structures that lend themselves somewhat readily to interpretation was based primarily on operational considerations. To reduce the effects of bias and rigidity introduced through operational expediency the players are required to interpret in detail the role they select.

In addition to the seven role-playing groups there are four umpires, each of whom has expertise relating to some aspect of the game. The umpires are responsible for postulating and stating as a fact of life the social, environmental, economic, and political consequences upon the total urban context of the developments instigated by the role-playing groups. In essence, the umpires describe the state of the game world.

The equipment required to play URBANISTA consists of a game board on which all developments are recorded, maps and a few forms. To facilitate the attainment of different objectives, several currencies or resource units are used in the game. The units are: Economic, representing capital; Populist, representing popular support; and Expertise, representing the expertise necessary to deal with problems of the quality of the urban and natural environments. These resource units are invested in the development of land, with each type of project requiring a particular combination of resource units. In addition, Populist Units are used as votes in the election of government officials, Expertise Units are used by the government in deciding which projects to implement, Economic Units are used in buying what money buys. There is no set rate of exchange between the resource units; in effect they are floating currencies subject to the particular pressures of the game.

URBANISTA is cyclical game in which the same basic operations are repeated in every round. Each round is divided into three phases and, in turn, these are divided into a series of events and activities. The first phase involves the selection of roles by individual players and the definition of objectives or goals by each team. At this stage the government expresses its objectives in the form of an official policy statement. During the second phase each team develops the strategy it thinks will best serve its objectives and translates it into a preliminary development proposal which is made public. Lobbying with the government and negotiations with other teams helps finalise plans into definite development proposals. The government approves only those proposals that are consistent with the official policy. The second phase is recycled three times in each round, making development a slow incremental process. The third phase is concerned with the evaluation of the consequences of development that are ruled by the Umpires but whose actual occurrence is determined by a combination of probability – initially assigned by the Umpires and later modified within certain constraints by all the players – and chance. At the conclusion of each round the Umpires report on the state of the simulated city. Returns on investments are directly related to this report.

URBANISTA is designed to permit participation by the players in determining the constraints of the game. This flexibility allows the game to adapt to the evolving situation and to the changing intentions and objectives of the players. It extends to each individual player who is free to change his personal values and objectives every round. Teams can also change in composition and redefine their objectives; governments can fall and new ones can be formed. It is possible for a player who was initially concerned with the effects of urban development on the existing wildlife in the area to become a politician, or for a politician to discover that making a profit is more rewarding to him than public service. It is also possible for a team that was initially concerned with education to become concerned with the plight of the poor.

The final output of the game is a physical design accompanied by a complete record of events and consequences. The data obtained in the course of the game is consolidated by writing it, in the same sequence in which it was derived, into a scenario. This scenario describes the step-by-step development of the final design. This design is the product of the combined values, objectives and strategies of all the game participants.

In the Urban Design Class where this game has been played the participants were students, guests and teaching staff. In this situation URBANISTA was utilized as a heuristic tool; however, if this game is to be utilized as a method for user participation in design it must involve the users in the gaming process. Involvement is critical to informed playing of the game as well as to the mutual education and understanding of all the interested parties. It should be noted that since URBANISTA is a game that may be applied to a variety of problems the particular factions described earlier were those utilized in the Halifax Urban Fringe version only. As a gaming framework, URBANISTA required that a representative cross-section of interest groups concerned with the particular problem be included in the game. However, the government and the umpires are included in every version of the game. For instance, if the design problem is concerned with a housing project the factions represented in the game may well be the future residents, developers, architects and the present residents of the neighbourhood in addition to the government and the umpires. Each faction can in turn be subdivided into groups; the future residents may be subdivided into housewives, young people and adult males.

URBANISTA is still under development and is likely to continue to evolve as experience of playing it is accumulated. However, some initial general observations about the overall utility of this method for achieving participation can be made. First, an important quality of such games is their ability to involve individuals who represent a wide spectrum of interest groups and points of view. Second, such games create interaction through what is, in essence, a carefully structured encounter. This structured encounter permits the interaction of divergent standards and values; trade-offs and compromises are made relative to the situation at a particular moment in the game. Third, gaming seems to generate enthusiastic and meaningful participation by the players and this is an important feature if this method is to be used as an approach to user participation. Finally, such gaming and simulation techniques allow for the reduction and simplification of the real world and its increasing complexity. This reduction of time and space in a simulation of a particular situation can result in considerable savings of money, material and perhaps men, as is the case in war games.

The modelled milieu of a game can in effect be a laboratory where ideas can be developed and tested prior to actual implementation and where the environmental designer can accumulate a factual foundation on which to base his ideas.

Gaming and simulation techniques seem to possess much usefulness, so far hardly tapped, for developing methods of bridging the gap between user and designer. However, gaming is only one possible approach. If participatory design is to reach its full potential a multitude of different approaches must be developed, each predicated on the right of each individual to influence the shaping of the environment about him.

OPERATIONAL GAMES APPLIED TO SOCIO-TECHNICAL PROBLEMS

Alberto Feo

Recently, in planning and other professional spheres, the word participation has been gaining ground very quickly. In every conference or paper, there is some concern for 'citizen participation'. To some, it means political organisation. To others, it is just a kind of relationship in a decision-making activity, and while 'total control' is advocated by some, a 'say in the decisions' is advocated by others. In these brief notes, I will not attempt to review the different meanings of the term because this would require a complete study in itself. Such a variety of interpretations however, suggests that there is a gradation or scale of participation.

I would suggest that participation is important in two ways. First of all, participation guarantees that some solution is reached in the implementation stage of any problem. Second, participation guarantees feedback from those who are affected by the decisions to those who decide. The first of the two participation levels is based on the assumption that actions are better implemented if those related with it directly or indirectly are conscious of the arguments for or against the actions, and as MacKenzie (1967) points out; "Significant changes in human behaviour can be brought about rapidly if the persons who are expected to change, participate in deciding what the change shall be and how it shall be made". This has also been called 'the participation hypothesis'.

The second level of participation is, curiously enough, the *corner-stone* of democracy (at least in principle), and I say curiously because, although the idea is applauded by everyone, "the applause is reduced to polite handclaps when this principle is advocated by the have-not Blacks, Mexican-Americans, Puerto Ricans, Indians, Eskimos and Whites. And when have-nots define participation as a redistribution of power, the consensus on the fundamental principle explodes into many shades of outright racial, ethnic, ideological and political opposition" (Arnstein, 1969).
Total democracy cannot exist if participation is not exercised at *all* levels of society and in all degrees. What does exist today in most countries (of both the East and the West) are 'degrees of sectional democracy'. *Degrees* because citizens participate only at the very bottom of the scale of intervention. *Sectional* because in most cases democracy is for 'someone' or for 'some group' of society and not for 'all'. Both facts defeat the principle of *total democracy* by definition.

THE PARTICIPATION ENMITY

As can be inferred from what I have said so far, the participation process is dependent on the relationship between the group which *has got control* and the group or groups who *want control*. This relationship has been expressed by Arnstein (1969) in the form of different degrees of participation which go from 'manipulation' to 'citizen control'. The problem arises when we have to create mechanisms to give to the power-less some power, and to take some from the powerholder. This, I suggest, can be done in two ways: one, by voluntary abdication of power by the powerholders; two, by force from the powerless groups. The first alternative is unlikely to occur and therefore we are left with the second course of action. Other factors, however, are present in this relationship between the powerholder and the powerless. Those factors can transform a relatively simple problem into a complex political issue and a democratic petition into a violent protest, and are those factors which we should study if we are to implement participation in any form.

Because participation is desirable to both people and government, it must be achieved in some way. But because it manifests itself in the power-interchange between two groups, any *action* must be seen in the light of the interests of each group. Unfortunately, until now, very little understanding has come from the powerholder about the demands of the powerless, creating, as a consequence, conflicts which can very often only be solved by force. Participation, if it is to be achieved without force, must come from the understanding of governments and other organisations that it is through people-involvement and commitment that problems can be solved and goals reached. If this is so then one should be able to call this, the *participation rapport* instead of the *participation enmity.*

ACTION AND PROTEST GROUPS

This study is concerned with the increase in the number of community groups which organise themselves to protest against some technological developments which affect in one way or another their environment and life in general. It is my conviction that most of these groups act in accordance with their beliefs and their honest concern with the effects of some technological developments which don't seem to take into account the social, moral and sometimes even the physiological aspects of those who are to be the users, or are to be affected by the developments' use and existence.

I believe that one way to find a solution to these types of problems is to increase the awareness, knowledge and intervention of communities in those issues which affect them, and in a way this study attempts to explore how this can be done.

Nowadays, problems are so complex and difficult that hardly anyone can see the whole system. This fragmentation in the appreciation of problems has also been reflected in the communication system between government, institutions, corporations and the people, the users and so on. This has had drastic effects in the feedback mechanisms of society, and, viewed from this angle, demonstrations and protest groups have become the 'natural substitute' for the traditional communication media, today closed to the people.

In the same way as the human body generates defences against virus and other 'invaders', society, in the face of a big anomaly, generates its own defence mechanism to prevent its degeneration. After all, it is through protest groups that history moves; it is through those who rebel against a corrupt, obsolete or unjust system that the great steps of history have been made. Governments and institutions forget too easily that many among those who are considered as heroes today were persecuted yesterday.

Protest and action groups should not be regarded as foreigners or intruders in the domains of society but as the natural consequence of an anomalous situation. Without them, very little would have been done in relation with noise in cities, the site for the third London Airport, pollution of the air and rivers, bad housing conditions of some sections of the population, the need for increases in wages, reductions in taxes, the need for public transport, and a thousand and one other issues.

Protest groups and action groups are only the primary living forms of what could, in the near future, be a complex and highly efficient system of control for our society, and they may be the key for our take-off towards a new society in the near future. They represent 'citizen power' and 'citizen control' in a very elemental form. Instead of seeing them as enemies we should see them as guardians of society. Protests and actions are the whole bases of social and political change. Our control systems should not be frightened by protest but encourage it. The people should be educated to provide constant feedback into the decision-making mechanisms.

DYNAMIC MODELS FOR SOCIAL CONFLICTS

The failure of traditional methods of dealing with social problems, especially socio-technical conflicts, calls for a new approach and new methodologies. The traditional methods of observation so often used in the social sciences, are not good enough to deal with complex problems where it is not an easy job to break down the problem into sub-problems. Instead, a dynamic and comprehensive technique is required; a technique which allows the scientist to study socio-technical conflicts within their context and which allows him also to include certain variables which until now have been ignored, mainly because of the difficulties involved in their manipulation.

How can we analyse social conflicts? This can be done, I suggest, by means of the so-called operational models; models which allow the user to represent in them different states, equivalent to situations in the real world, and by changes in some of the parameters of the model to simulate specific situations.

We can identify four types of operational models:
1) Business Models,
2) Rigid Computer Models,
3) Game Theory,
4) Operational Games.

The one specific type of operational model in which we are interested here, and one which promises to be of great help in the study and solution of socio-technical conflicts, is the so-called operational gaming.

One of the main advantages of using operational gaming is the fact that, unlike most of the other models (especially those in which computers are involved), it admits irrational behaviour into the model and therefore accepts inconsistent factors as part of the process.

AN EXPLORATION OF STRATEGIES IN COMMUNITY CONFLICT SITUATIONS

My research has been directed towards the exploration of the use of operational models, especially games and simulations, in the analysis and development of strategies when facing specific socio-technical conflicts and problems.

We limited ourselves to those problems which have local effects on the communities. Problems at a national scale would require greater time and resources, which we do not possess. We also limited our study to the development of strategies by the affected people, be they communities, action groups, local authorities, etc.

The main aims of this study can be summarised as follows:
1) To explore the use of operational models in the development and evaluation of alternative strategies in conflict situations between communities and other organisations when dealing with socio-technical problems.
2) To provide a tool which would allow action groups to explore a range of alternative futures when dealing with socio-technical problems at community level.

To achieve the stated objectives and to explore some areas of interest, it was decided that a simulation of a real situation (which has already happened), at a level of community conflict which would not require much technical data, was one of the best ways in which the feasibility of the use of operational models could be assessed.

The Simulation

Basically the 'simulator' consisted of two groups which would be in conflict. One group, the Community; the other group an Institution, a Department of Government, a Local Authority or any other organisation. There was a third group which represented the media through which information could be transmitted, i.e. Radio, Television, Telephone, Newspapers, Mail. A further group was that which controlled the functioning of all this.

The Problems

Two conflict situations have been used as bases for the simulation experiments. Both problems were real, and the information, names, dates, areas, etc. are all true and had been taken from the original records of the sessions and meetings of the community in which the conflict took place. Both problems were in areas of low income levels, although housing conditions and environment were rather different in each case.

The problems had been recorded and analysed by Action Groups which worked in the areas, and thanks to them we have been able to obtain such information.

The first problem related to the existence of a ship canal (Rochdale Canal) which goes across the north-east of Manchester carrying problems not only in terms of smells, flies, rats, and many other things, but also in terms of the increasing number of children drowning in recent years. The inhabitants of the area have been trying for some years to convince Local Authorities of the need to make the area safe, both in terms of health and human lives. Many schemes have been proposed and all sorts of meetings and protest marches have been carried out. However, the canal is still there and children continue to drown.

The second conflict situation was located in Hattersley, a community to the east of Manchester, and of fairly recent construction. In Hattersley there is basically one main area in which most activities are carried out, where there is a church, shops, town centre, school, etc. The Local Authority decided to expand Stockport Road, at the moment a minor road through the area, to convert it into a four-or-six-lane motorway-type road. The people of Hattersley then decided to talk to the Local Authorities to try and make them change the route of the road to avoid the division of the community into two areas, on one side the facilities, and on the other, most of the people.

The Experimental Sessions

Each session has lasted for nearly three hours, including 'warming-up' and 'debriefing'. There was no inflexible time limit, though we asked the groups to try and solve the problems in a maximum of three hours. Some times this took a short time and some others significantly longer. There were two sessions for each problem-group.

After every session, players were asked to fill in a questionnaire and a diary of their impressions, problems and recommendations. Complete on-line records of all communications both within and between groups were kept, using written records, audio-tapes and video-tapes during the sessions.

Conclusions

Our experience suggests that people without the explicit knowledge and experience of a socio-technical problem can be helpful in the exploration of such problems. This should not be interpreted as if they are 'the best' subjects for this kind of exercise, because much better and also more difficult than using 'any' people is using *'the'* people involved in the real problem or situation.

One of the most obvious uses in which we see this kind of technique being of use is in the area of community work or community organisation or community action groups. It may also be helpful in solving conflict situations different to the ones we have mentioned, e.g. solution of industrial strikes, in planning as an evaluation tool of policies, etc. To the question: "have we provided a tool for community groups which would allow them to explore alternative futures when dealing with socio-technical problems?", the answer should be – "Not yet". However, there is plenty of evidence that these kinds of models, generated and modified by continuous revision, can be structured. We only scratched the surface of the real problem, and much work in this direction is still needed before the first concrete results can be seen. Meanwhile, I would try to apply these techniques to a real community and to monitor and record the changes and results from actions and conflicts.

ACKNOWLEDGEMENTS

This paper is an abstract from a dissertation by the author, prepared at the Design Research Laboratory, University of Manchester Institute of Science and Technology (Feo, 1970). The study was carried out under the sponsorship of the Consejo de Desarrollo Cientifico of the Universidad Central de Venezuela.

INFORMATION PROCESSES FOR PARTICIPATORY DESIGN

Yona Friedman

At the roots of any scientific method there are some very important constraints of an epistemological nature. They concern the subject matter the method deals with, the tools it uses, and finally what and how we know of this subject matter by using the specific tools. These constraints define the method by setting its limits.

It follows directly that if the subject matter (thus implicitly the limits) of a scientific method suffers quantitative change, then the method as a whole has to be transformed. It happened this way in physics, in mathematics, in biology, and it is happening in behavioural sciences – architecture and planning included.

The classical method used in architecture was a simple chain of operations, which started with the future user of the architect's product (we will call him 'user' or 'client'). This client had some specific needs, and he explained them personally to the architect. The architect made a plan translating the specific needs of the client into a hardware object supposed to satisfy these needs. This plan was realised by skilled artisans, and when the hardware object was finished, everybody was content in the best of the possible worlds.

I have parodied somewhat this process, only to underline the architect's role as the 'translator' of the client's specific needs into the language understood by the skilled artisan. Thus the architect was a necessary person only in cases where the client had no common language with the artisan: at no stage of the process could there be any doubt that all decisions had to be made by the client himself.

At first sight, the situation today looks exactly the same, except for one new fact: the number of clients has become very large. But alas, our initial considerations imply (and it has happened in reality), that this one new fact was sufficient to transform completely the original situation.

In our primitive image the client simply told the architect about his specific needs. As, very often, these needs were not at all explicit, the architect had to spend quite a long time getting sufficiently informed about the client's needs. As an example, I could cite a very famous architect, who stated publicly that "it takes generally about six months to understand the client's way of life". Then how long would he work to understand 10,000 clients?

Obviously, the question is a rhetorical one. It would take a time longer than all written history of mankind refers to. But, for any architect or planner or designer to work for 10,000 clients is actually not at all unusual.

There are but two solutions to this kind of problem:
1) to produce so many architects (or planners) that there are only a few clients left for each architect (or planner),
2) to shorten the information period.

The first solution would evidently transform architecture and planning into the largest profession that ever existed. This solution seems rather unlikely, and unimplementable. Logically enough, the second solution was the one the profession followed. Unfortunately, however, it was followed in possibly the most absurd way. Let us examine why.

Architects (and planners) thought thus: "We cannot determine the specific needs of each individual future user, so let us determine the average needs of each individual future user, or to put it another way, the specific needs of the average future user".

I don't think that I need to explain in detail what were the results: there is massive discontent manifested by all individual users of architects' products. The reason for this discontent is obvious: *the average user is a non-existent one!* If there are satisfied but the average client's needs, it is logically implied that *no specific needs of any individual user can be sufficiently satisfied.* Thus, we satisfy the non-existent client instead of satisfying the existent one.

This reasoning is obviously simplified in order to state the situation. The real situation is far worse: in the above I admitted hypothetically that the architect (or the planner, or the sociologist) is capable of determining objectively an average. In reality, not only the chain of thought was fallacious, but the estimation of this average was inevitably falsified, because of the preconceived ideas of the person doing this estimation.

This crisis situation is the result of the profession ignoring the fact that the clients or users (the *real* decision makers) became a very large number, and thus that the whole process the architect is participating in has to be transformed.

THE INFORMATION CIRCUIT BETWEEN USER AND PLANNER

Let us reconsider the situation. In the traditional process the mechanism of the process was as in Figure 1. The architect (planner) and the artisan were nothing but the 'channel' by which the 'information content' (or 'message': specific needs) was transmitted to the resulting hardware.

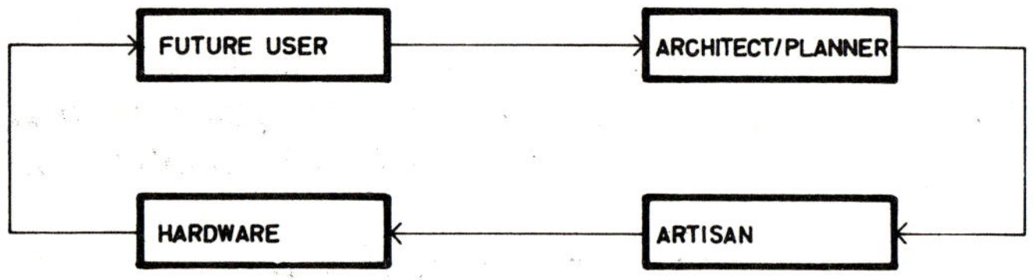

Figure 1

The process was a simple one, composed of a 'transmitting station' (future user), a 'channel' (architect and artisan), a 'receiving station' (final hardware) and a direct 'feedback' (usability of the hardware). The system did not permit any correction or adjustment as the result of an unsatisfactory feedback. Thus, if the 'receiving station' (house) did not get the 'message' (specific needs) from the 'transmitting station' (client), the responsibility lay with the 'channel' (architect + artisan). Because no adjustment was possible, the 'testing period' (the above mentioned information period of the architect) for the 'channel' (architect) was very long, and thus no, or little adjustment was necessary once the work was finished.

Once this scheme had been transformed for a growing number of future users, the system became fundamentally different, as Figure 2 shows.

Figure 2 represents the 'ideal' situation, as the architect and the planner imagined it. Let us suppose, only as an hypothesis, that it could work. The diagram contains two 'bottlenecks': at the stage of the architect and at the hardware stage. The architect's bottleneck is narrow for the incoming information, the hardware bottleneck is narrow for the outgoing feedback. The system's vulnerable points are: the architect's handling of information, and the hardware's adjustment for varying individual use.

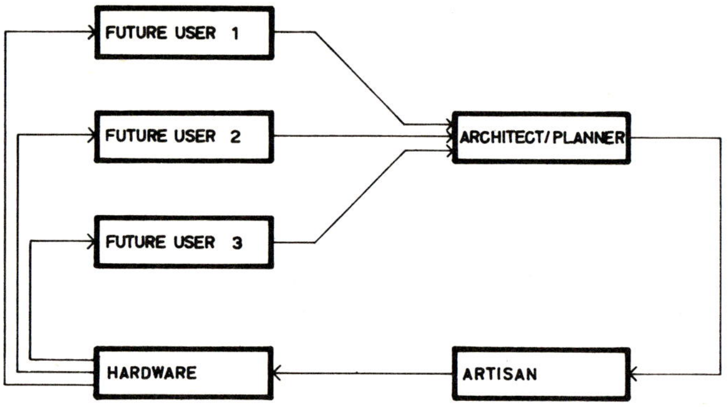

Figure 2

The 'average man' concept evidently does not improve this situation. On the contrary, it adds but a new stage (where errors are possible) without eliminating the bottlenecks. The same bottlenecks stay, even if one is in a different place (Figure 3). I call these two bottlenecks the 'information shortcut' of the architect and the planner.

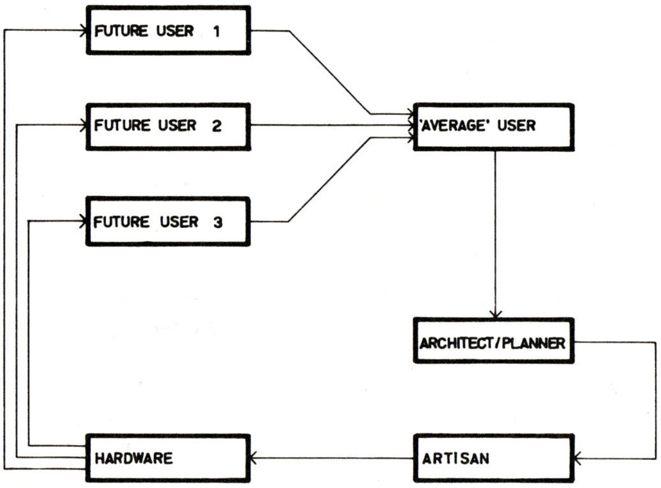

Figure 3

LOOPING THE USER'S MESSAGE DIRECTLY TO HIMSELF

Any science, discipline or system is essentially based on information. It is critical to a discipline the information manipulation it does in a given context. Obviously enough, when the context changes, the manipulation of information has to change with it. We found above that a part of the information manipulation (the 'channel') in the architectural process did not change when the context changed, and the result was that the process had to adjust itself at another stage: the use-adjustment of the hardware. Unfortunately, this adjustment was physically impossible. This is the current crisis of planning disciplines.

Our purpose will be to construct a new process, eliminating the 'information shortcut', and in consequence the unreliability of the message. Such a process would be like Figure 4.

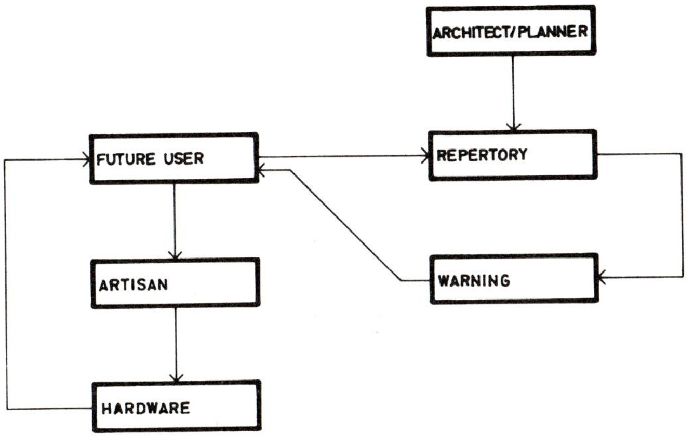

Figure 4

This diagram (Figure 4) can be simplified to the coupling of two loops, both without any 'bottleneck', as in Figure 5. The first loop represents the act of decision made directly by the future user. This loop does not contain any other 'station' than the user himself and a repertory containing *all* physically possible combinations of the organisation realisable in the hardware. The second loop contains uniquely the user and the hardware.

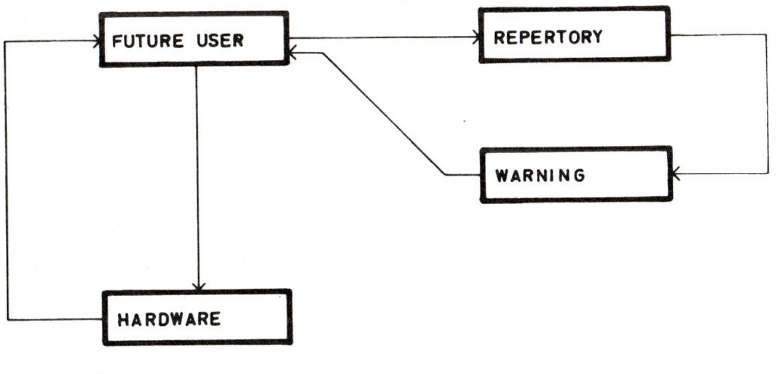

Figure 5

In both loops the intermediary person of the 'plan maker' (architect, planner) is eliminated. The plan maker could be included in the second loop, if desired, as a translator between the client and the hardware, but surely not in the first loop. The possibility of a bottleneck in Figure 4 is eliminated by breaking the original loop into two separate ones. This operation permits us to arrange side by side as many double loops as required (as many as the number of users might be), because all loops stay distinct all their length, as shown in Figure 6.

Let us explain the significance of this figure, through an imaginary but implementable example. In such a scheme the future user (client) meets, instead of the architect, a *repertory of all possible organisations* (solutions) that could be implied by his own individual way of using the future hardware. This repertory, (which is necessarily a finite one) has to be presented to him in a form he is capable of reading. For each item in this repertory, additional to the notation of each solution, there is an associated 'warning'. This warning informs the future user, again in terms understandable to him, about the utilisation issues (advantages and disadvantages) created by himself for himself, by choosing this specific item from the repertory. (The 'warning' is not based on any particular value system, but on intrinsic logical properties of the solution: thus it can happen that the same 'warning' represents an advantage for one user and a disadvantage for another, as the two might have very different ways of life.)

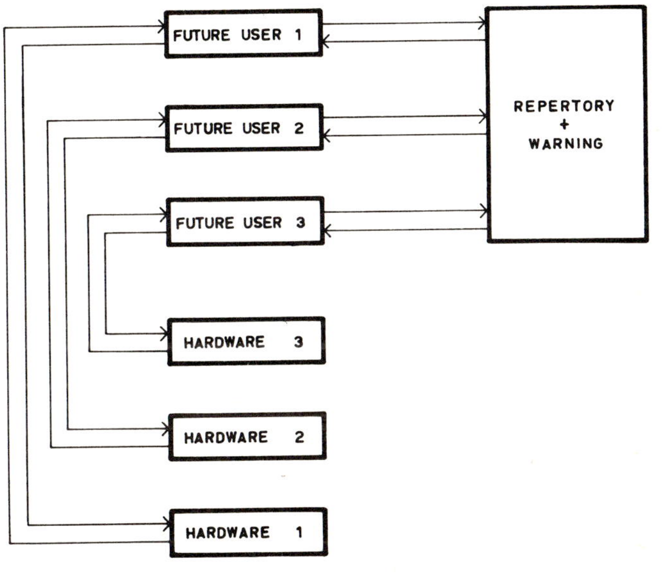

Figure 6

Thus, the first loop is the future user's *self-information* by the user of a repertory. The first loop ends by the future user actually choosing one of the items listed in the repertory.

The second loop is very similar to the classical process: the future user communicates his choice directly to the artisan producing the hardware. For this communication he will use the code of the repertory item corresponding to his choice. The artisan (or industry) produces the hardware chosen by the client, and the client starts to use the self-chosen hardware. He will take all responsibility in the process, because he was warned in time concerning the properties of the solution he chose.

Both loops in this redesigned process begin and end with the user: the first loop is that of the decision, and the second one that of the communication of the decision (implementation of the decision). In both loops there is no specialised 'channel', no translator necessary.

At first sight it seems that the architect or planner, who has kept for centuries the role of the translator, is eliminated from the process. Reality is somewhat different. The 'channel' is not eliminated in the redesigned process: it is the repertory itself, or, to be more exact, it is the *notation* used in the repertory. This notation (mapping) has to be understandable by any user as well as by the artisan (industry) constructing the hardware. Thus the new process returns to the scheme in Figure 1, where there are three elements: the future user, the 'channel' (in Figure 1 the architect, in Figure 6 the repertory), and the final hardware product. It is not the architect or planner who gets eliminated from the process, it is but their old role. There is a place for them with a *new role: to construct the repertory*. The professional man – who constructs both the repertory and the warning – has to be a scientist: a 're-designed' architect and planner.

DEMOCRATISATION OF DESIGN

Both the future user and the planner-scientist have to know the same things: how to read a repertory and how to choose within a repertory, and how to read a 'warning'. This knowledge should be taught in the primary school. The planning scientist will learn more than this primary school knowledge. He will study the objective rules, and their implementation in constructing the repertory and the warnings.

There is nothing unusual in such a situation – mathematicians and laymen use the same operations and rules on different levels, and anyone can read a meteorological chart although only the scientist can construct it. They key to the well-functioning of a democratically regulated environment is in this two-fold learning.

It is the future user to whom decision making legitimately belongs. Hence my concern with 'democratisation', because democracy means that everyone has his individual part to play in decision making.

Decision making involves risks to be taken by the decision maker. Any system that does not assign the decision making to those who will have to take the risk resulting from an inappropriate decision is an immoral system. Yet such is the system as practised by architects and planners: they take the decisions, and the future users take the risks.

THE RESTAURANT EXAMPLE

Restaurants, all the world over, have their services organised on the basis of a repertory. Their repertory is what we call a 'menu'.

The menu contains all dishes (thus all pre-existing elements) the restaurant has in stock. Each dish presented by its name in the menu often has there as well a summary explanation of how this particular dish was prepared, and it is assigned a price: we will call these statements, associated to each item in the list, a 'warning'. The role of the 'warning' is to inform the customer (the user of the list) about some of the consequences of choosing certain elements from the list: these consequences, in the restaurant example, might be economic ones (concerning the customer's purse) or health ones (concerning the customer's stomach), etc. The customer is well informed, if only he reads the menu.

The customer, after absorbing the information, will compose for himself a combination of the listed dishes – a meal. There are possibly no two customers who will choose literally the same combination, even if they take their meal together – a meal is dependent on personal taste and choice. There is no better or worse composition, if meals are concerned; no general value table.

The restaurant owner will not interfere with any choice made by any customer, even if this latter's choice does not conform to his own (the owner's) personal taste, or even if he considers this choice as more than extravagant. His task is to serve the customer with his choice based on the repertory and the warnings contained in it. The restaurant owner's 'creative' work is the construction of his 'menu'.

This example shows the way I propose for the construction of the repertory serving the customer of the architect or planner. This repertory will be based on the complete list of possible space divisions, linkages and labelings (mappings of the problem itself). The client (we called him before the 'future user') will have the practical freedom to choose any possible assemblage, without having to follow in an obligatory way any other person's (the architect's or the planner's) preferences. The repertory contains, beside each possible linkage, a 'warning' corresponding to its 'effort economy'. This 'warning' informs the client of the advantages and disadvantages of this particular linkage for the client's particular use-pattern (i.e. about the 'effort costs' which are intrinsic properties of each linkage).

The last task incumbent on the architect and the planner (always following the restaurant example) is to have 'in stock' all the linkages (or the corresponding hardware) a client might choose, and, once the client has made his choice, to give him the possibility to alter his choice or to correct it (and so the corresponding hardware) once he finds it desirable to do so.

The restaurant example covers completely the architect's and the planner's problems. If there is an additional point neglected by the restaurant owner but not negligible for the planner, this is in the fact that the architect's and planner's 'warning' has to be directed not only to the particular client who actually makes the choice, but also to the community (in the form of some 'carbon copy'). The community has to be warned about the immediate consequences each individual choice might imply for it.

THE NEED FOR AN INFRASTRUCTURE

The architect or planner's 'menu' therefore contains a number of possible pre-existing space divisions, which may be linked together (with access-points) in a number of possible ways (the number of possible combinations available to the future user will therefore be large, but finite). Also, for each item listed in the repertory (i.e. the 'menu') there is a particular 'warning'. This warning can have two components: the first will be simply the purchase price, the second will show a characteristic of what the purchaser can expect from his choice. This characteristic can be called the 'utilisation efficiency'.

The 'utilisation efficiency' can be computed from a utilisation matrix of the frequency of use of each linkage which the future user will make by using the assembly of enclosures in his particular way. Thus can be constructed an 'effort value' representing our 'warning' system. It warns the future user of the implicit consequences of using in his particular way the assembly of his choice.

All the preceding material, however, has concerned but the act of choice: who has to effect it, why, what is a repertory, who is to construct the repertory, what are the issues implied by choice of a determined alternative? The next question is: what is the hardware which contains the possibility of constructing any one of the combinations in the repertory, without any exception?

Let us look at the character of such a 'non-determined' (non-committed) hardware. Such a hardware becomes necessary because construction operations cannot wait for a final decision from all future users. We have to find a type of hardware constructable before all the choices are made, and capable of supporting any enclosure organisation corresponding to any individual choice, whatever it may be. Such a hardware type we call an 'infrastructure'.

There are two possible ways of constructing this infrastructure. Either it can consist of an initial set of totally unconnected spaces, and the user's choice implies cutting access ways between the spaces in order to link them into the chosen assembly, or it can be a skeleton, and the user's choice implies constructing separating walls to form the space enclosures.

However, there must always remain the possibility of the user 'correcting' his choice once the hardware is constructed to suit his initial choice. This implies that everything within the infrastrusture (thus every operation of either 'cutting through' or 'separating') should be reversible. Thus all the corresponding hardware should be mobile – only the infrastructure itself can be rigid. If I could give a metaphor, it would be that only the infrastructure should be 'inked in', and all the linkages should be 'drawn in pencil'. The correcting action, to use this metaphor, would be partial erasion of the pencil lines by a rubber. Thus the fundamental hardware conclusion is the physical separation of the fixed infrastructure from the mobile in-fillings.

AN EXAMPLE OF IMPLEMENTATION: THE FLATWRITER

As an example of a practical implementation of this theory, I will cite the proposal I prepared originally for the World Exposition 1970 in Osaka. The 'Flatwriter' is a machine with which any future user of a building can write down his personal preferences as for his own environment, and do this in a 'visual language' which an architect as well as any future users of the same construction can understand. The machine contains a repertory of several million environmental arrangements, with warnings about issues implied by each individual act of choice, both for the individual future user and for the community.

The Flatwriter is based on the separation of four information circuits. The common point of these circuits is the person of the future user, who is the only decision maker in the process, as shown in Figure 7.

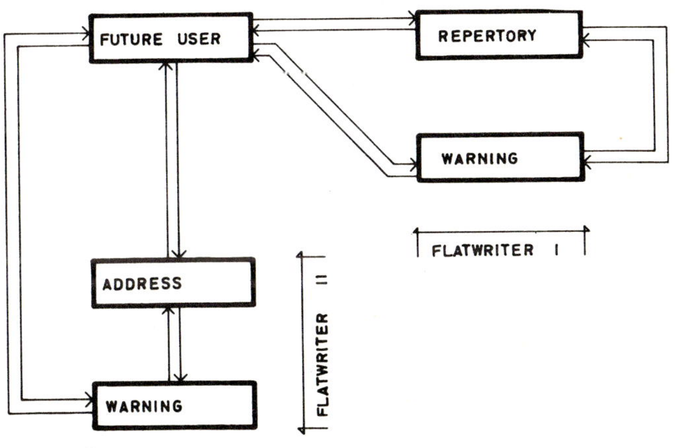

Figure 7

Let us follow step-by-step the operations of the 'mechanism' constituted by the future user and the Flatwriter.

1) The future user finds himself before a keyboard containing four categories of signs: a) configurations of volumes, b) shapes of individual rooms as admitted by the contextual building system, c) type and situation of services equipment (kitchen, bathroom, lavatory), d) orientation of the proposed flat. The keyboard contains in each category the corresponding complete list of possibilities, and the client has to choose, following his own preferences. He chooses in the following order. First, he will push one of the configuration keys; then three keys for choosing the shape of the first, second and third volumes; then, in the same consecutive order, for the equipment for each volume in the desired position; finally, he chooses the orientation of his self-designed flat. Thus he pushes a total of eight keys, writing an eight letter 'word' for a three room flat. The Flatwriter has 53 keys, and can write down about two hundred million different flats. The machine's *raison d'être* is conveyed by this enormous number of terms in the complete list of three room flats. Such a large list could not be meaningfully presented, but the elements and their rules of composition can be.

2) The Flatwriter I part of the machine prints the chosen plan, as any honest typewriter does. A copy will be kept by the client (future user), and a coded record will be forwarded to the computer.

3) The client gives his choice of finishes for the flat, using a coded catalogue of technical variants. This catalogue would vary in different local contexts.

4) Flatwriter I computes the price.

5) The client gives his estimated frequencies of using the different parts of his flat.

6) Flatwriter I calculates and prints the corresponding 'local effort' diagram, based on the client's chosen plan and utilisation matrix.

7) The client chooses his preferred address within a given infrastructure (multi-storey, skeleton megastructure).

8) The Flatwriter II part checks whether the chosen site is free, and does not obstruct access or light for a previous choice made by some other client. If the choice is alright, it registers the new choice in the memory, and prints the chosen co-ordinates for the client. If the choice is not admissable, a signal instructs the client to choose again.

9) Flatwriter II updates the settlement's 'local effort' diagram, calculated on the basis of the previous individual choices. It forwards a print of the new local effort values belonging to the newly-chosen site and its immediate neighbourhood to the client.

10) Flatwriter II presents the corrected 'local effort' diagram on a screen for the whole community. The corrected diagram is superimposed on the previous one, showing thus all changes (advantages and disadvantages) caused by the last individual's choice.

These ten steps roughly cover all the decision making processes which happen in the planning and forming of a settlement and community. The steps 1, 3, 5 and 7 are uniquely dependent on the client's decision; steps 2, 4, and 6 (printing and warning) are made by Flatwriter I (circuits taking the paper and pencil task of the project design as traditionally done by the architect); steps 8, 9 and 10 are controls effected by Flatwriter II (circuits performing the planning authorities' bureaucratic tasks).

The Flatwriter has, like the ordinary typewriter it was inspired by, an enormous repertory. Using the Flatwriter, decision making is the result of a co-operation between man and machine. There is a constant dialogue between the client and the Flatwriter. The adjustment process (choice of preferences) is the client's reserved domain, and a further dialogue is possible between the individual client and his community.

ADAPTIVE-CONDITIONAL ARCHITECTURE

Charles M. Eastman

An intimate relationship exists between man's pattern of activities and the environments he has built to surround them. Most of us have access to a kitchen that provides hot and cold water and heat for cooking at a convenient level above the floor. Usually the person doing most of the cooking sees to it that the arrangement of knives, cutting boards, and the storage of cooking equipment in the kitchen are organised in response to the procedures of cooking. Our air-conditioned, artificially lit, and sound-insulated buildings are obvious attestations to the fact that man has the capacity to modify almost every aspect of the physical environment to support his activities, and that he also has an almost infinite ability to adapt himself to them.

The relationship between human activities and the surrounding environment results from adaptations in both directions. Man has always adapted his own behaviour to his surroundings, for example in the amount of clothes he wears or the adapting of work to fit the tools he has available. Architecture and engineering focus on the adaptation; they adapt the surroundings to desired human behaviour. While both kinds of adaptation will take place any time a human uses a space, the ethic of modern design is to take human activities as given, without constraints, and to create an environment which maximally supports them. Instead of constantly adjusting his own actions to meet the structure of the environment, a person in a well designed environment is free to act with the environment fully supporting him. Freeing man from the constraints imposed by the environment has been one influence allowing the surge of creative power evidenced in western material technology.

THE ARCHITECT'S CONCEPT OF FIT

We assume it to be a requisite if man is to evolve higher social technologies also. In general, the relation between an activity and its environment has come to be called its *fit*.

I wish to specify the fitness relationship between activities and an environment precisely. Given some pattern of activities, I use fit to designate the relative amount of effort required (in physical, psychological, social, or economic terms) to carry out those activities in a particular environment. The less effort required to carry out the activities, the better the fit. In this sense, fit is a measure of the degree to which activities are unconstrained by the physical environment.

In general, fit defines the relation between one pattern of activities and one environment. For each pattern of activities proposed for a space, there would be an individual measure of its fit. The measurement of fit includes as components the physical effort and time required. In these areas, human factors, ergonomic studies, time and motion studies and circulation analysis are all partial measures of fit. The fitness of an environment encompasses the breadth of social and psychological influences also. Proxemic influences and social interactions, plus the psychological influences of sensory processing, cognition, and symbolic references all play a role in determining the fit of a human activity to a space (Figure 1).

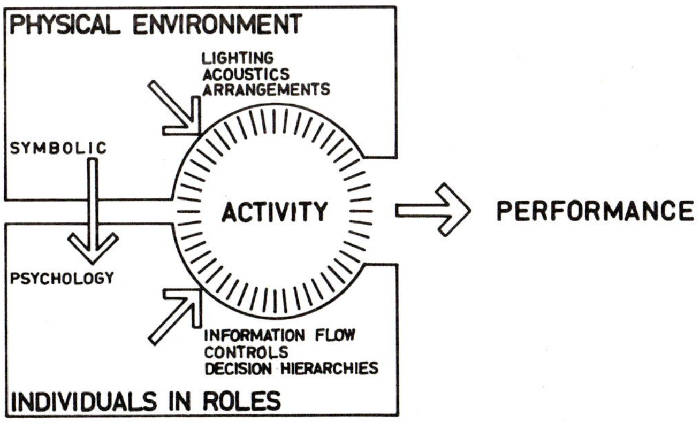

Figure 1

By now I would expect all of us to agree that fit is an important design concept. Whether or not it is considered as explicitly as I have presented it, the concern for fit is part of the design of all entities which are used by man. Almost the sole purpose of the design brief is to define good fit in a particular design situation.

In this paper I would like to focus on three neglected aspects of the concept of fit. These aspects are of

growing importance in the types of architecture we will increasingly build, now and in the future. The three aspects of fit are:
1) measuring fit when users are an anonymous group,
2) predicting fit in a new unrealised environment when all existing behaviour is constrained by existing environments,
3) controlling fit between activities and space when the activities change over time.

After explicating these three aspects of fit, I shall describe an approach to design called *Adaptive-Conditional Architecture* which allows significant resolution of the three issues. Adaptive-Conditional Architecture derives from cybernetics and control theory and focuses on the dynamic and continuously changing aspects of the physical environment rather than its static and monumental aspects. Lastly, I shall present two introductory examples of design resulting from the application of Adaptive-Conditional Architecture, and outline the processes required if our environments are to become more adaptive.

MEASURING FIT FOR ANONYMOUS USERS

Architects are generally competent in designing an environment for a single user. They use their creative abilities to manipulate form to achieve an environment that strongly supports the user's preference and behaviour. This is possible because the architect knows how to ascertain the behaviour and activities of that user via personal interviews. The architect's traditional modus operandi is closely akin to anthropological methods of observation and informal analysis. The architect traditionally has relied on a key informant and his own empathising.

This approach is less than satisfactory when multiple anonymous users exist and especially when they have different cultural and behavioural patterns from those of the architect. In this context, the architect's own value system and intuition will lead him to false conclusions and result in the imposition of *his* values on the users. In the U.S.A. this situation has led to serious problems in the policies of urban renewal. Middle class values are being imposed on many people, while destroying their past friendship and socialising patterns. It would be like building new towns in England without any pubs. The point is that architects do not have familiarity with techniques for gaining design inputs from large numbers of users, and deriving from large numbers of inputs useful and responsive design criteria. Obviously, greater understanding of the techniques of sociology is required.

PREDICTING FIT IN NEW ENVIRONMENTS

The measurement of fit for multiple users is well understood by others, if not by designers. The second aspect of fit is not well understood and derives from the adaptation between behaviour and the physical environment.

Any observation of activities or behaviour must be made in an existing environment. That environment influences and in some way constrains the existing behaviour within it. Change the environment and the behaviour will change also. Thus simple extrapolations from behaviour in existing spaces to new spaces is likely to lead to many unanticipated problems. This difficulty has been anticipated in proxemic studies and some interesting new approaches are being tried.

Similarly, direct questioning or interviewing is not likely to lead to the creation of environments without constraints either. People are often aware of the most constraining aspects of their environment, but not of the more subtle ones. Thus one step forward is possible from direct work with a user, but this will only result in another, albeit possibly less crucial, set of constraints.

CONTROLLING FIT FOR ACTIVITIES WHICH CHANGE OVER TIME

Related to the above issue is one pertaining to the evolution of activities. In removing the constraints of an existing environment, the designer may significantly improve the fit between the activities in the old environment and the new space. The confounding issue is that the activities in the new space will not be the same as in the old space. Both sets of activities quickly adapt to their space. Similarly, activities constantly change due to exogenous influences. Production procedures, organisational structures, treatment in health care, teaching methods in school, life patterns of the family are but a few examples of the changes imposed by wider social influences which affect the fit between activities and space.

AN ANALOGY

The three aspects of fit that have been neglected are easily explained through a simple analogy. This analogy also points out the major thrust of Adaptive-Conditional Architecture. I shall consider the aspects of fit in reverse order.

The analogy I wish to use is that of a heating system for a house. All such systems, whether electric, gas, or coal, produce heat by the conversion of fuel. The input is electricity or an oxidising fuel and oxygen. All can be adjusted for a variety of conditions. The fit of the heating system is the degree that it provides temperatures within the home suiting its occupants.

Now the third neglected aspect of fit is similar to the design of a heating system where the heating man puts the equipment together, samples the weather and number of occupants of the house, and adjusts the heat for proper level, then leaves. Such a heating system will work well until the first change in conditions. The valves to adjust the system are there but require manual operation.

Our architecture today has the same degree of adaptability. It is tuned prior to occupancy, but is not responsive to internal nor external changes after the facility is in use. All operations require the intervention of an expert. Fit is only considered during initial design and not during use. The fit is static and cannot respond to change.

Modern heating systems, though, require no intervention. The thermostat provides automatic adaptation of the heat output to the changing conditions of the house. The same possibility exists for other aspects of our buildings. Circulation, lighting, the arrangement of rooms all can become more automatic in providing fit between activities and surroundings.

The second neglected aspect of fit corresponds to the design of an improved heating system for a very cold climate. If previous heating systems could only provide heat to achieve 50°, people would be wearing heavy clothes inside. If we size the heating system by measuring the amount of heat required for the comfort of its users, we can predict that the heating system will be sized too small. The changes in behaviour resulting from removing their heavy clothes should be anticipated. Again the thermostat, with its ability to respond to a range of conditions, coupled with a general heating system that can provide varying amounts of heat, is the solution. Similarly in buildings, singular changes in arrangement promote changes in behaviour. The frequency of use for a coffee room in an office is a function of its accessibility. If close, meetings and other business-oriented activities may begin to move there. Empirical evidence from the facility will not predict such changes, but an adaptive environment may respond and support them once they become manifest.

The first neglected aspect of fit corresponds to a heating system designed by one person for himself only and without adjustments. It is likely that the system would not suit others. Large empirical studies have led to good knowledge of the range of temperatures desired by different people, and we solve the variation issue by allowing individual control of most heating systems. Heaters allow for differences with a manual control setting and good heaters allow control not of the heat output but of the desired temperature. The physical variable which corresponds one-to-one with the individual's comfort thus can be directly controlled.

This analogy suggests that greater fit between human activities and architecture is possible in a dynamic and evolutionary setting if architecture incorporates automatic feedback mechanisms corresponding to the thermostats for heating systems. I call architecture providing such automatic adaptation Adaptive-Conditional Architecture.

By now, some obvious shortcomings to the current approach to architecture should be apparent. We currently design most buildings as if the activities within them and the situation outside them are stable and unchanging. We attempt to achieve fit for the life of a building at construction time. We walk away and return, if hired, to make adjustments, e.g., remodelling, at infrequent intervals and at high cost.

On the whole, five years after the completion of a building, the fit between it and its activities is lousy. I suggest this is the original designer's fault. The means for adaptation have not been considered.

ELEMENTS OF ADAPTIVE CONTROL MECHANISMS FOR ARCHITECTURE

Let us look more closely at the thermostat mechanism. Its elements may provide us with guide-lines as to what is needed to realise an Adaptive-Conditional Architecture. In Figure 2 is presented a standard

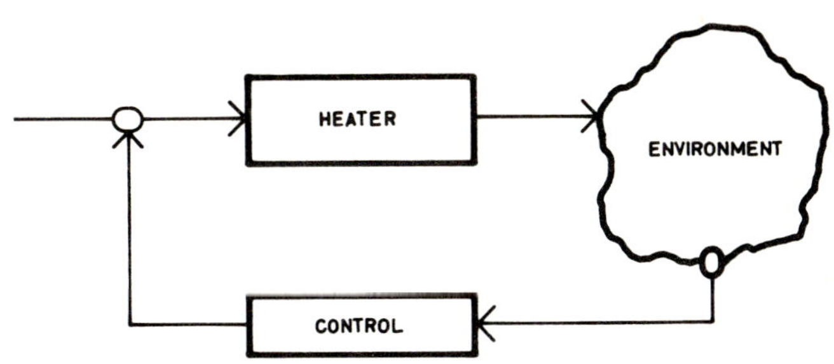

Figure 2

heating system, as described earlier. The thermostat mechanism is shown below the heating device and consists of four elements. First, a thermostat involves a *sensing device* for temperature of the space. Temperature is taken as the critical variable of the heating system, i.e. that dimension in which goals may be defined. The sensing device returns a signal to the *decision algorithm* which is the second element of the thermostat. The decision algorithm defines what

decision is to be made for any range of input of the critical variables. For the heater the algorithm is a set of two simple conditional statements:
IF TEMP ⩽ X *AND* HEATER = OFF *THEN* ON
IF TEMP ⩾ Y *AND* HEATER = ON *THEN* OFF
Most heaters have such a two-parameter decision rule. By adjusting the difference between X and Y, the rate of cycling can be easily controlled.

The third element of the thermostat is the *change mechanism.* It receives the signal from the decision algorithm and carries out any actions. In the heater, the change mechanism is a switch or valve. The last aspect of the thermostat is its *control setting* feature. The algorithm incorporates a means by which a user can define his own setting of the critical variables. In this way, individual variations are accommodated.

In Figure 3 I propose the corresponding elements of an Adaptive-Conditional Architecture. Here the entity being monitored by the control mechanism is

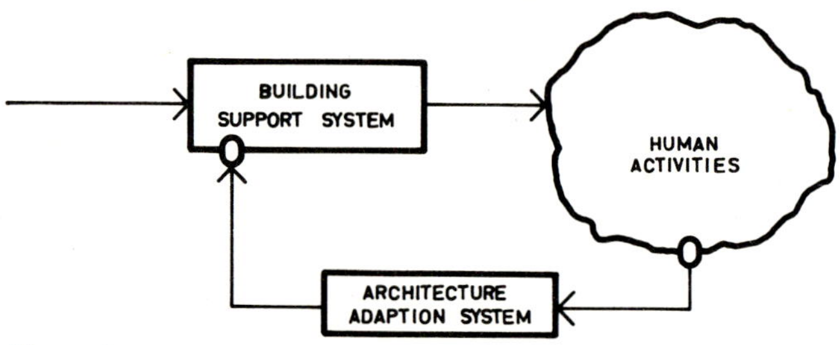

Figure 3

a building. The building provides spaces and material support for human activities. Although the activities are affected by things other than architecture, it surely is one influence operating over time. When the building does not support the activities as it might the architectural control and feedback process should provide beneficial adaptations. When new occupants take the space they should be able to reset the goal variables to achieve the environment that best suits them. With such an adaptive mechanism, the need to accurately predict the exact form of environmental support of users when designing space is not required. Their evolving behaviour can be expressed by changing the setting on the adaptive mechanism. *I am arguing for the development of the software techniques that will allow the hardware of architecture to become continuously and automatically responsive to users.* The primary difference between the thermostat and adaptive architectural controls is that heat is a flow and is continuously being produced. A building is an entity (in economics, a stock) and thus is not being continuously produced. While a thermostat changes a flow, an adaptive architecture must change a stock, a much more complicated process.

It is not necessary to incorporate all four elements of the feedback mechanism in order to provide Adaptive-Conditional Architecture. Adaptive-Conditional Architecture can be considered as a continuum moving from zero adaptability to infinite amounts. The elements of adaptation provide a taxonomy for examining examples of design already approaching Adaptive-Conditional Architecture and for inventing new techniques which promote its ends.

TWO EXAMPLES OF ADAPTIVE-CONDITIONAL ARCHITECTURE

One set of critical variables of a space or building's effectiveness in supporting a pattern of activities pertains to horizontal circulation. For offices, schools, and hospitals, it has been shown that good layouts of activities and circulation make the organisation more efficient. One study, for example, reported that the staff of an operating suite in a hospital spent 23% of their working time in walking. A more efficient circulation system was generated which was calculated would lower this to 17%. The 6% of time saved amounted to about 12,000 dollars per year in salaries (or productivity). Thus the fit between an organisation's pattern of activities and the arrangement of space and circulation can save time and money, and make work less fatiguing.

In order to generate efficient layouts of large numbers of rooms, it has become common to use computer programs for space allocation. They take as input a circulation matrix, as shown in Figure 4.

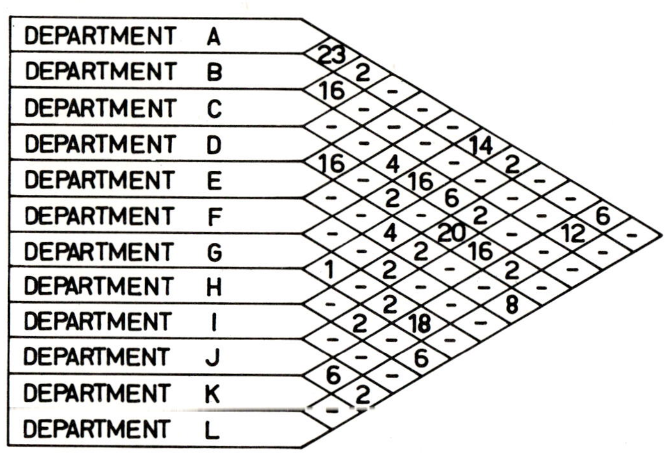

Figure 4 Interaction matrix.

Each entry in the matrix represents the number of trips between two departments, over some time period, such as a day. The total distance travelled between two departments is the distance between them times the number of trips taken. Thus the sum of all circulation is the sum of the distance between each pair of departments times the number of trips between them. The computer programs arrange room layouts and select one that provides a low value to the total circulation function. Many companies' independent recognition of the efficiency problem resulting from layout has made "office landscape" a popular means for easily changing use of space. Office landscape was first developed in Germany and is finding increasing use in the U.S.A. Instead of fixed partitions, office landscape relies on bookcases and other furniture to separate different spaces. The whole arrangement can be easily changed. An example floorplan derived from the office landscape concept is shown in Figure 5.

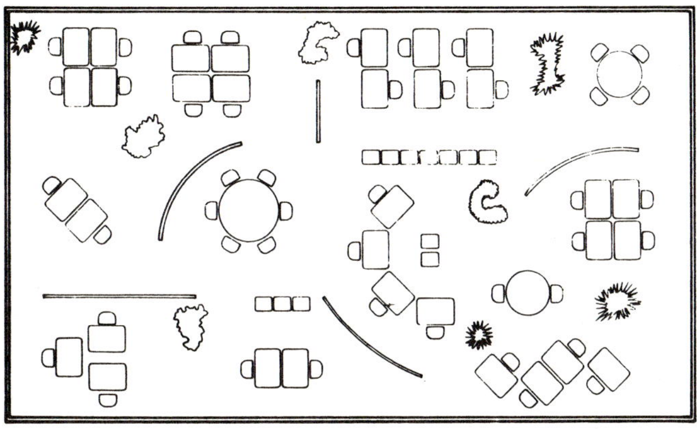

Figure 5 Open plan office.

Adapting the office landscape to evolutionary changes in operations is carried out in the following way. Interior planners or management consultants with architectural training come to a company yearly and carry out the refitting of the environment to the organisation. They accomplish their job by sending a questionnaire to each user to record his own circulation for a day. Certain location preferences, e.g. near windows, special lighting, or other fixed attributes are also identified on this questionnaire. From all the returned questionnaires, the consultants generate a trip matrix and run it through a computer program that determines an efficient arrangement that satisfies individual preferences also. Department relocations are suggested.

The office is changed over a weekend and work continues. Office landscape fits well into the paradigm of Adaptive-Conditional Architecture. The critical variable is distance walked. The sensing device for measuring the critical variable is the questionnaire. The decision algorithm is the computer program. The change mechanisms are the easily relocated equipment coupled with the weekend remodelling. The location preferences ascertained by the questionnaire allow for setting controls, e.g. individual preferences.

Besides arrangement and circulation, office landscape techniques analyse lighting, communication procedures, and equipment requests, and other aspects of the environment which support activities. The approach could be expanded further, to include acoustic controls, special floor and wall surface needs, visibility needs for certain kinds of supervision and monitoring, etc. Such expansions are likely to require new types of equipment with special design characteristics. Adaptive-Conditional Architecture has important form implications.

My other example of Adaptive-Conditional Architecture deals not with the adaptation of a facility after it is built, but on its adaptation while it is being built. The need for such adaptation exists in the planning of community facilities at the neighbourhood level. Such facilities may be special purpose, such as health care or a library, or they may be multi-purpose. They may be provided by a private or public organisation.

In a large number of cases where such facilities have been planned in the United States, conflicts have arisen because new needs have been identified by members of the community, but the planning and design process has not been able to accommodate them. The problem arises as a result of the large need for public services in many communities; as soon as people start thinking of one, they begin thinking of others that are needed also. The learning process of the community allows it to think of opportunities only slowly. In the typical situation, a debate between the providing institution and members of the community ensues when some new function is proposed that cannot be incorporated because of the cost of changing an already largely fixed plan.

The problem is that architectural adaptation during planning is currently too expensive. Lacking are the sensing devices for identifying critical variables that are likely to change during design and afterwards, effective decision processes for responding to change, and mechanisms for implementing changes.

A procedure for better tapping community priorities and changes in them has been developed by the author and his colleagues. At the neighbourhood level, we proposed a citizen information system which at regular intervals would provide an overall sensing of critical social, public service, and attitudinal variables. The variables that have been proposed for collection in a trial application in two public housing projects include security, accident hazards, maintenance, use of space, social concerns of the neighbourhood, social services, plus others. These general indicators provide quick feedback "pointers" to potential problem areas. When a particular goal variable suggests a problem, then more detailed interviews, public meetings and data gathering is undertaken to determine both the cause of the problem and to identify alternative means for rectifying it.

The information system provides another important contribution in the form of a common data base for determining priorities. By providing the public with the results of the monitoring process, they will gain a wider perspective of the community situation, not only from their own limited experience, but also from the experiences of their neighbours. Idiosyncratic concerns will be minimised while emphasising those that are common.

Our applications to date emphasise public services as well as physical facilities. We have begun examining how public housing management, for example, may begin to respond to the information received from such an information system, in terms of management policy and operating and capital resource expenditures. These methods have obvious implications in the design of new facilities. As neighbourhoods change over time, or simply as people age, we expect that such a system will allow more effective provision of services and facilities than are now possible with ad hoc methods.

A direct implication of the development of an Adaptive-Conditional Architecture exists for architectural education. In the desire to provide orderly design problems in our schools, teachers take fixed and often arbitrarily defined problems. Yet development of design procedures allowing programme changes during design would provide a social benefit for the future profession. Design problems for public facilities where use is likely to change during the design phase are a reality. As an example, I would propose a problem for a neighbourhood health centre. The week prior to presentation, the brief could be modified to include a teen recreation centre in the evenings. This kind of flexibility is non-existent in our design education and in our profession.

THEORETICAL AND TECHNICAL ISSUES IN THE DEVELOPMENT OF ADAPTIVE-CONDITIONAL ARCHITECTURE

Earlier, four elements of Adaptive-Conditional Architecture were identified. They were:
1) a sensing device, for monitoring critical variables of the environment;
2) a control algorithm for determining appropriate actions for each state of the environment;
3) a change mechanism for implementing adaptations identified by the control algorithm;
4) a control setting feature allowing input of individual preferences.

In this section we review each of these elements in detail and suggest areas where research or new technology is needed. Several possible innovations suggest themselves as a result of the adapting conditional approach. As a general remark, to develop mechanisms to automatically adapt the environment to new conditions over time requires a much more precise explication of a design problem and its objectives than is now necessary. The dimensions of a design which are important, their interrelationship, and their range of desired values for all anticipated conditions are all necessary if a dynamic fit is to be achieved between activities and the environment.

The first element of an adaptive mechanism for architecture is a sensing device which monitors critical variables. A major issue concerns which parts of a facility are critical to the performance of activities. No answer to this question will universally suffice. A theory is needed which allows one to define, albeit open-endedly, the function of a building.

While such a theory remains to be fully developed, the outline of such a theory suggests itself directly from the previously described premises of Adaptive-Conditional Architecture. Adaptive-Conditional Architecture considers a facility as an activity support system. Any functional environment (e.g. architecture, not sculpture) is designed for the purpose of facilitating certain activities and behaviour for a group of people. The range of behaviours and users should be specified for a facility, if these are delimited. In general, though, any activity can be considered as dependent for environmental support along the following dimensions:
1) an area of appropriate size and shape to hold an activity. The arrangement of areas for different activities should also facilitate sequences of or interaction between activities;
2) environmental supports – temperature, lighting, air quality – for the activity;
3) areas and supports for equipment used for activities, from furniture to a computer;
4) appropriate barriers – sound, security, dirt, bacteria and animals – between different kinds of activities and between public and private spaces;
5) information access and displays necessary for the activity;
6) explicit identification of user group.

This listing is a conservative one, not explicitly dealing with symbolic or aesthetic issues. Each of these support dimensions is expected to require a somewhat different setting for different personalities and social groups.

In order to develop appropriate sensing devices for each of these areas of support, appropriate physical measures must be developed. While many unresolved issues exist in measures for all five areas, the existence of measures of any kind are lacking in the last two realms. Yet information display is clearly one of the dimensions in which the environment contributes and which changes most quickly.

Sensing devices for Adaptive-Conditional Architecture range from periodic surveys or interviews to observations by specialists, to built-in sensing units. Even where adaptive techniques are already utilised, advances are possible from more automated sensing devices. An example is the automatic sensing of circulation in a facility, allowing re-allocation of spaces whenever the transition cost is offset by the resulting performance increase. Sensing devices in the areas of lighting and acoustics are possibilities today, yet they have not been tried. The predominance of fluorescent fixtures with their fixed output has discouraged adaptive techniques, though wattage savings due to adjustability of other types will partially offset their cost. The introduction of "white noise" as acoustical background to a space is increasingly common. Adaptation of the level of white noise required to mask the current level of background sounds is a possibility not yet explored, to my knowledge.

Compared to the simple thermostat, control mechanisms for an Adaptive-Conditional Architecture are bound to be quite complicated. One example of a control system allowing adaptation of an environment as use varies is the elevator control system.

Control mechanisms take as input the goal parameters by which the effectiveness of some aspect of the environment can be measured. For acoustics one goal parameter may be reverberation time, another the decay rate of different sound frequencies. Circulation, on the other hand, includes frequency of travel, the value of the time taken for that trip, cost of distribution utilities to different parts of the building, and the costs of controlling air quality, sound, or other pollutants that are created by some activities and undesirable for others. The output of a control mechanism used in Adaptive-Conditional Architecture is an action to be taken with some respect to some part of the environment – to move a wall or activity space, to alter the qualities of some materials. The algorithms for making these decisions are likely to be the same that the designer would use if he was originally designing the facility. Of course the objective is to develop techniques whereby buildings are continuously and automatically being designed and redesigned over time.

Strict technological problems exist in the creation of change mechanisms that will allow adaptation of physical environments. We do have examples – gymnasia which convert into auditoria, stadia that function for several sports, possibly with a moveable top. The choice of whether adaptive change mechanisms are to be manual or automated depends on the relative costs of the two kinds. If a change is expected to be required every two months and the costs are one hundred dollars for manual labour versus 50,000 dollars capital investment for automated change mechanisms, it is not likely that the change will be automated; the payback period will be over fifty years. Similarly, the rate of change and its value is the estimated worth of the change, in efficiency of productivity, versus the cost of the change. Changes will be made more often the less expensive they are. Cost-benefit analysis is the tool for making such decisions.

Change is traditionally made to a space just prior to its occupancy; this is the rule for offices, medical clinics, and houses. After occupancy, change becomes much more expensive. The costs of interrupting operations for a long enough period to make the changes is more expensive than the benefits that would result from the changes. Changes must be made less expensive.

In the thermostat, the control setting is the substitute for user studies. The control setting allows the user to bring his values into the decision-making process. In order to achieve a similar arrangement in physical design, we must first: a) identify the critical variables that allow variation according to individual preference; b) develop the appropriate control algorithms to allow this input and to alter its outputs so as to reflect the inputs of the user; c) develop the input mechanism by which the user communicates his preference. This may be a mechanical control, as for lighting or acoustics, or it may be a survey or questionnaire that allows identification of more complex but less quickly changing preferences.

CONCLUSION

I have attempted to show that architects today take only limited consideration of the fit between the environments they design and the activities that go on within them. If the fit is to remain good, then the building must become continuously adaptive to the activities within it. Our analogy was the control and adaptation process used in the heater.

If buildings are to provide continuous fit, then their design must focus on the control processes allowing adaptation. Needed are sensing devices, control algorithms, change mechanisms and control setting features. By focusing on these mechanisms and designing them along with a facility, we designers may guarantee that the fit of the environment is improved.

SELF ORGANISING ENVIRONMENTS

Sean Wellesley-Miller

It is almost a platitude to say that the city makes man as much as man makes the city . . . or did, because in a direct formative sense we no longer do so; architects and city planners do it for us. A predominant feature of the architectural scene ever since the emergence of total urbanisation has been the subjection of ever larger numbers of people to an urbanisation process in which they play little or no formative role. We sometimes tend to forget that this is a relatively recent development. Until a century ago only some 10% to 20% of the built environment was consciously planned (Burnskill, 1970) – the remaining 80% to 90% being socially-evolved: that is, constructed along traditional lines without the help of a professional town-planner or architect. In many so-called primitive societies, which however often possess complex and contextually well integrated architectures, the professions of architect or planner are unknown even today.

In fact, socially-evolved architecture, with or without artisan specialisation, was the dominant mode of environmental formation even in the West up until the industrial revolution. Within the industrial context the socially-evolved process broke down – or at least failed to find new procedures relevant to and viable in the changed environmental conditions. After a period of *laissez-faire*, the socially-evolved process was abandoned in favour of a professionally planned environment.

As a result of this change we now have a situation in most western countries in which some 70% to 80% of the built environment is consciously planned, although considerably less of it may be architect designed.

This change from a socially-evolved to a professionally planned environment is probably one of the most significant events in the history of architecture to date. In any case I believe it to be very relevant to our present concern with user participation in design, and with this in mind I would like to take a closer look at these two processes, and especially the former, from a more formal point of view.

THE ON-LINE/OFF-LINE DISTINCTION

The distinction between on-line and off-line systems is fairly well known in computer technology and process control, but since I will be using these terms in an unfamiliar context it might be best to state explicitly what is meant by them here.

Conventionally, operations carried out under direct computer control are termed 'on-line'. However, I will say that an on-line system is any system in which the operations of control, evaluation, decision and execution are carried out at the same rate at which the system itself operates. Driving a car through traffic, or a football game may serve as examples.

Conversely, by an off-line system I mean one in which this is not the case. An example could be production control in a factory. Here we collect data, construct a flow diagram of the operations involved and run the resulting simulation through a computer, changing the parameters until we arrive at some desired state of affairs. We then return to the factory to implement the changes indicated by our results. The computer may be miles away from the factory, the time taken to process our algorithms has nothing to do with the time taken to perform the equivalent operations in the factory and we can return to implement our decisions at will. The control system is off-line.

Now, my contention is that socially-evolved architecture is an on-line process which can be abstractly represented by a diagram (Figure 1) taken from

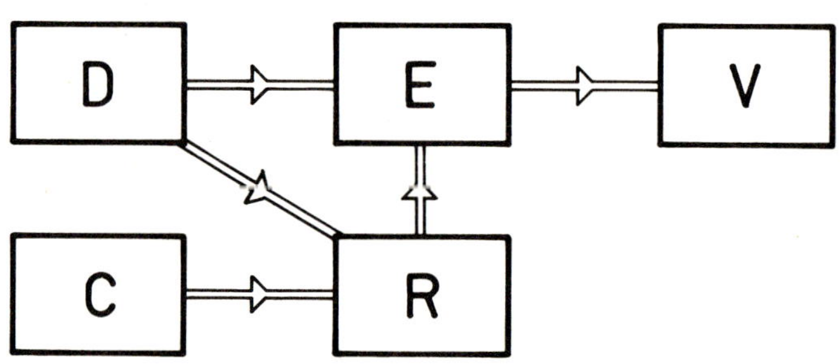

Figure 1

Ashby (1956). The arrows in the diagram represent communication channels. The block D represents a vector of environmental disturbances; C a control function (tradition) that constrains R's actions. R is the inhabitant who receives information from both D and C and uses it to regulate E (environmental circumstances – architecture) in such a way as to hold the value of certain essential variables V, which may be social as well as climatic, within acceptable limits.

Similarly, a professionally-designed environment is an off-line process which can be represented by Figure 2 (Willis, 1970). A crucial feature of an on-line process is the importance of the time parameter, that requires all events to take place in real time. The formation of a socially-evolved environment is a time-dependent or 'historical' process, with all that this signifies in terms of involvement and meaning. This is something that tends to be completely absent from professionally designed environments as far as its future occupants are concerned. Consequently, even if the structure of the environment arrived at by each design process proved to be identical, the on-line process would almost certainly invest the environment with more 'meaning', simply because of the mode of its formation.

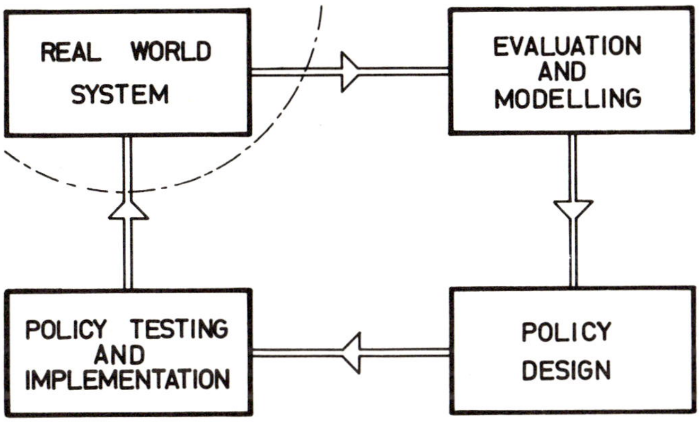

Figure 2

STRATEGY AND TACTICS

The business of design, however it is achieved, takes place in an increasingly complex, often reactive and in recent years increasingly changeful reality. During the lifetime of a building built today, say some fifty years, many features of the social, technological and even physical environment in which it is designed to function will have changed. Yet it is extremely difficult for us to forecast with any certainty when, how and which of these factors will change. This raises some very difficult problems for the designer. In searching for possible solutions I will turn to the work of Schützenberger (1954), and I will take the liberty of using his arguments and examples freely.

Schützenberger's first example involves a man on top of a hill who wishes to get to a house in the valley below in the shortest possible time. Between him and the house are many causes of delay such as boulders, escarpments and marshes which make travelling in a bee-line out of the question. An exhaustive and final solution to his problem would be for him to make a map of the district, divide it into small areas, find the time taken to cross each area, join the areas into all possible chains linking the hill top and the house and then find the path that gives the smallest total time for the journey. This path is the best possible and Schützenberger calls its selection in this way a strategy. Another and more common method would be for the man to select a point about fifty metres lower down, which he could reach rapidly, go to it and then to select another even lower, and so on until he reaches the house. Schützenberger calls this method a simple tactic and goes on to say that "the tactic differs from the strategy in that the tactic does not take into account the whole situation but proceeds according to a criterion of optimality that is applied locally, stage by stage". Closer examination of these two methods shows that the first is in fact a model-forming (the map of the district), off-line process, while in the second Schützenberger's walker is operating in an on-line mode, interacting with his environment and adapting his policy as he goes along.

Schützenberger continues his analysis of the relation between strategy and tactic by introducing the notions of a 'span of foresight' or planning horizon and a 'degree of flexibility'. The span of foresight is determined by that part of the total situation or subset of elementary areas the man can scan at any particular moment. In our tactical example it was 50 metres. The degree of flexibility is measured by the minimal time at which a provisional goal may be replaced by another. When the span of foresight covers the whole situation then flexibility is unnecessary and tactic and strategy are coincident. In a well-defined environment a tactical solution is usually inferior to the strategic solution, although in many instances they will be quite close to each other. One is reminded of Simon's advocacy of a 'satisficing' as opposed to 'optimising' criterion. Further, the cost, as measured by the amount of computation required to generate a tactical solution is mostly way below that involved in getting the best possible or strategic solution ($n(2^{n-1} - 1)$ compared with $(n - 1)n!$ operations in a square matrix problem with n rows or columns).

However, and of great interest to us as designers, Schützenberger demonstrates that if the environment is stochastic then the optimal strategy is just the simple tactic of trying to do one's best on a purely local basis. For example, the marsh may be difficult to cross after rain, a bull may or may not be in a field, a boat may or may not be available to cross a river and so on, and if our walker has no knowledge of the exact state of affairs before a particular trip, then he can only assign a probability to the duration of crossing a particular elementary area and the simple tactic becomes his best strategy. Another and well known illustrative example is that of a dog which wants to run to its master who is walking steadily in a definite direction. If the dog could calculate, its strategy would be to integrate the relative velocities and go directly to the point of intersection. Since it cannot do this then its best tactic would be to run towards the continuously varying point where its master is at, so following a curved path. The tactic is simpler than the strategy but results in a longer path. However, if its master is moving in a completely random manner, so that the dog is uncertain where he will be at the next moment, then it can be proved that the dog's tactic is the best one possible.

The examples I have used are geometrical but there is no reason why they should be. The problem of designing a dwelling for a family whose future membership, social status, pattern of living and economic level is unknown but open to change is of the same sort. This is a real problem for the design of low cost housing in developing countries. A definitive solution embodied in a 'hard' design that will remain valid for the complete life span of the building is impossible. What we have to search for is a 'soft', adaptable design that can be made to respond over a short period of time to the changing needs of its occupants.

Not only in developing countries but also in the West the future environment is becoming increasingly complex and probabilistic; like our dog we can no longer predict the future trajectory of the system with certainty and, like the dog, our best bet might be a tactical design which aims continuously at the observable state of affairs rather than trying to design a complete strategic solution. In other words, for technical and economic reasons we may have no option but to design in an on-line, participatory mode.

ON-LINE TACTICAL DESIGN

The old constructional vocabulary of brick, tile and timber, individually articulated and reworkable in use, has given way to a new one of prefabricated components of concrete, steel and glass. Wall and beam have been replaced by slab and skeletal constructions; the local artisan by large contractors using industrialised methods. In the course of these changes, environmental control has passed out of the hands of the local community into that of governmental authorities, professional designers and large contractors. In many ways the new materials, methods and building forms reflect these latter people's requirements more than they do those of the occupants – as anybody who has tried to bang a nail into a concrete wall will appreciate.

Obviously, design participation is not simply a social or organisational problem; it is also a technical problem involving materials, structural systems and construction techniques. Design participation does not stop at the construction stage but continues throughout the functional life of the building. If taken seriously it has definite consequences for the quality and structure of the social system/physical structure interface, and for how the built environment looks and functions. For example, while it might appear desirable and democratic to ask the future occupant during the design process how he would like his house laid out, there is a certain naivety in this. Houses last a long time – and do *you* know how you will want to live ten years from now? Options must remain open. It must be possible to recover from mistakes. At present a building is designed, built, used and eventually demolished. In the future all four activities may tend to continue throughout the life time of the building. The constructional equipment would be designed as an integral part of the building. The structural system would have to allow for extension, upgrading and adaption. Design would be continuous. The building would cease to be a fixed object and become a facility.

As an example of a design problem approached in an on-line mode, I would like to take the problem of designing a park. Let us imagine that instead of staring at our drawing board in the hope of inspiration we went to the site and scattered some seeds around, placed some saplings in pots at random and went home leaving the gate open. Some of the seeds would fail to grow, others would get trampled on but some would take root and grow. Similarly, some of the saplings would be destroyed and others moved, but after a time the situation would be more or less stable. Paths would emerge: a winding path up the hill made by people wanting to take advantage of the view from the top and continuing, say, to a picnic spot on the sunny, lee side, or meandering off into the grove of trees. Anyone who has studied natural paths over dunes or through a wood will know how sensitively placed yet resistant to replacement these can be.

Instead of a six-week design project on the drawing board we would have a participatory design process that might last anything from two to five years on the ground; the park would have an 'historical' development that was the result of many peoples accumulative actions. The job of the designer would be to stimulate, steer and stabilise the process. He would pave the path up the hill so that old people and mothers with prams could navigate it without difficulty. He might place a rest bank on a bend in the path, and a telescope on top of the hill. He could place refuse bins by the favourite picnic spots and even benches and a table at one or two of them. If the process failed to get going, or subsequently flagged, he might indulge in a bit of bulldozering to make things more interesting, or introduce some open water. Since the process would work with the stream of natural events, the maintenance costs of the park, once it had reached a stable state, would probably be much less than those of a normal park.

The next example is based on my experiences in the Nieuwmarkt area of Amsterdam – an old workers' quarter built in the seventeenth and eighteenth centuries and now embedded in the centre of the city. Currently, new interlocal roads are being built through this area in a rather destructive manner. Whether these roads are in fact necessary is a debatable question, but if they are then it is due to a particularly atrocious piece of off-line planning on the outskirts of Amsterdam called Bijlmermeer – a dormitory estate housing some 20,000 people. This plan has been on the boards for fifteen years. Its possible consequences for the Nieuwmarkt area and other parts of Amsterdam are fairly obvious. Assuming that the roads are indeed necessary then work could have begun on them at least ten years ago. Instead of demolishing a broad band of houses through the area, old sub-standard houses could have been weeded out piecemeal here and there and the empty lots filled with playgrounds, gardens, parking lots and short term buildings with appropriate functions. This would have allowed the area to adapt slowly to the emerging trajectory; the road would have grown out of, rather than having been imposed on the existing structure of the community. We ran a simulation of such a process on an IBM 360 and came up with a number of viable results, although such a simulation cannot be used in place of the actual process. I know that this is not a very impressive example of the on-line process but I have included it just because of its everyday nature.

These examples have been more in the nature of speculative proposals than concrete design examples, so I would also like to discuss two very different but actual on-line design processes.

The first is the architecture of the Marsh Arabs in Southern Iraq (Salim, 1962). I think this is especially interesting because the built environment has emerged during the last seventy years or so. Previous to this date the Beni Isad were a tribe of camel-herding bedouins who lived in the desert. They were driven into the marshes, which are nearly as large as Holland, as a result of feuding. A more extreme change of physical environment would be hard to imagine.

The principle building material is the reed which grows in the immediate vicinity. Reed is also used for practically everything else: household utensils, cattle fodder, fuel and the artificial islands themselves are built from it. The lifetime of a *sarifa* or *mudhif* is about seven and twelve years respectively. They require constant repair. Environmental and social circumstances change very slowly indeed. The professions of architect, planner and even constructor are unknown. A man is dependent on his relations for help in building; nobody will work for money.

Rates of change are slow while reaction times are quick. Building materials are strictly limited but easy to work and determine the structural possibilities. A strong constraint is placed on radical innovation due to the social organisation. Design advances take place piecemeal as a result of constant interaction involved in repair and rebuilding. None-the-less a new building type has been developed within the last fifteen years. The layout on individual islands consists of a fluid and variable complex of sleeping platforms, storage spaces, work areas, winter quarters and a guest house. No two islands are the same; each one has been organised in response to family requirements and its position in the village. Over the years the layout of a particular island can change quite radically.

The second example deals with events on the other side of the world and in a completely different cultural context, yet, despite the huge differences, there may be something to learn from the comparison. This example is the 'Barriada' land invasion and self-building process which goes on in the urban environment of modern Lima.

More than 40% of Lima's two million population live in self-built barriadas, which were and still are for many people the only possibility for home ownership. Immigrants come to Lima in a steady stream from all parts of Peru. The reasons are much the same but the actual process is very different from that in Iraq. Most new arrivals either find temporary lodgings in the *cajahornes* of central Lima or stay with relations who have already established themselves. After having found some sort of employment, and saved, with great difficulty, a small sum of money, fifty or more immigrants will form an association and try to invade empty land on the outskirts of the city. If the invasion is successful, and sometimes it is not, then a rough plan is made on the ground complete with lots for a school, clinic, market centre, etc., and work begins on the houses. Most of these start out as reed mat shanties but may end up as a large three-storey building of brick and concrete. Materials are constantly upgraded and the buildings enlarged and reconfigurated over a period of years.

It is easy to misunderstand or romanticise such a process. Invasion makes the subsequent provision of main drainage, water and electricity much more expensive. Many buildings remain sub-standard and most of them could be improved upon by a professional designer or contractor. Nevertheless such a process does have definite advantages over a state planned development; among them Turner (1968) places the freedom of community self-selection, the freedom to budget one's own resources and the freedom to shape one's own environment. The barriada movement is based on a hard-boiled realism – if an invasion subsequently fails to progress quick enough people are quick to move on. However, the barriadas do show that an on-line participatory process based on a marginal economy can work in a modern urban environment.

On-line self-organising processes such as these deserve more study than they have received. Combined with a measure of planning they provide a possible answer to the urgent problem of shelter in countries where the need is great and design and planning expertise, and the facilities to back it, are in short supply. I believe also that such processes are not altogether irrelevant to our problems in the affluent West.

A PARTICIPATORY ARCHITECTURE

Perhaps our most urgent questions are: what does a participatory architecture involve, and how do we design it – for design it we must. It is unrealistic to expect an overall design to emerge using on-line techniques alone within the present context. I hope I have clarified the first question enough to allow us to approach the second.

Ashby (1956) has shown that "only variety can destroy variety". Thus we can only try to find or place some constraint on or in the vector of disturbance D (Figure 1) or increase the variety available to R, or, expressed symbolically:

$$H(D) + H_D(R) \leqslant H(R) + H(V)$$

or

$$H(V) \geqslant H(D) + H_D(R) - H(R)$$

where H is a measure of entropy and $H_D(R)$ means the uncertainty in R when D's state is known.

Adaptive environments employ the second strategem. A number of proposals have been presented along these lines, including Archigram's 'Plug-in City' and Yona Friedman's space grids. They all have, to a greater or lesser extent, one thing in common; rather than furnishing a 'hard' finished design they present a structured possibility-space allowing indeterminate development within controlled parameters. In the place of designing finished objects or structures we design systems or environments in which structure becomes equipment and equipment is responsive to variable needs.
The other possibility of placing constraints on the vector of disturbances can be seen in current proposals (e.g. Fuller, Otto) for huge domes spanning several kilometres and enclosing whole cities.

My own preference, since I rather like some wind and rain now and then, is to replace the technological megalomania of vast domes with an assembly of climatic envelopes on the community or zonal scale, forming a variety of covered and exposed spaces for various purposes. The interesting thing about climatic envelopes is that if you did encapsulate a city you would immediately make every building within the envelope technically obsolete. Wind and snow loads, rainfall and thermal contraction and expansion could be virtually ignored in the interior. This would allow us to design in a completely different way. Malleable, lightweight materials such as fabrics, impregnated cartons, foams, plastics and metal alloys could be used in conjunction with modern epoxy and sulphide glues and space-structural techniques. Living facilities could be continuously designed and adapted by their occupants using pre-fabricated kits, replaceable units or site-formed foams.

If the community spaces were air-conditioned, then a lot of functions which are now enclosed purely for climatic reasons could take place in the 'open air'. We would move towards a 'polynesian' architecture. Such an environment would also stimulate social changes in forms of association and individual life styles. Given these conditions, a soft tactical architecture incorporating user participation in both initial design and subsequent functioning could become a practical proposition without losing the benefits of modern technology. Whether it would find social acceptance is another question.

ASPECTS OF LIVING IN AN ARCHITECTURE MACHINE

Nicholas Negroponte

MASSACHUSETTS INSTITUTE OF TECHNOLOGY
SCHOOL OF ARCHITECTURE AND PLANNING
77 MASSACHUSETTS AVENUE, CAMBRIDGE, MASSACHUSETTS 02139

DEPARTMENT OF ARCHITECTURE
Room 9-518
Nov 15, 1971

Nigel Cross
The Open University
Walton Hall, Walton
Bletchley, Buckinghamshire
England

Dear Nigel,

Your letter of October 1st indicates that you have forgotten that I gave you a draft of "Aspects of Living in an Architecture Machine" on the condition that it not be used for publication. This was because the document is a very first draft of new and controversial (for our Group) ideas, presented in "draft" format, not in the King's English.

While my inclination is to insist that you do not publish it, I am willing to agree to its inclusion in your proceedings on the conditions that it remain as is (including the missing references) and that a photocopy of this letter be published as a preface. The paper is being reworked at this time as a chapter in a new book I am doing. A proper reworking of the paper, however, will not be available within the next few months and I cannot expect you to hold up your publication.

I would like to point out that all too often papers are presented and printed in a finished format that leaves the reader with very little idea of the vacillations and deadends that were integral parts of arriving at the conclusions, whatever they may be. To this end, I volunteer "Aspects of Living in an Architecture Machine" as a first stab at disclosing a train of thought. You are publishing our very first inklings of what the subject is all about and I am sure that our own feelings will change.

You should know that the actual title will be: <u>Ménage à trois</u>.

Sincerely,

Nicholas

Nicholas Negroponte
Assistant Professor of Architecture

P.S. The missing number on page four is "32", which says something about byte-oriented machines.

TELEPHONE UNIVERSITY 4-6900

Two apparently unrelated movements have marked the development of the theory of architecture in the past five years – that of participation and that of computation. In the one case, we are talking about providing 'users' of physical environments with a higher and more direct level of input to the criteria for design as well as to the design itself. In the other case, we are dealing with computers that aid the 'designer'. One obvious convergence for these two growing lines of research occurs by making the 'user' in participation be the 'designer' in computation (see proposals of the Architecture Machine Group, 1971). But that is not what this paper is about.

This paper is about another, not so obvious, and rarely taken seriously, concurrence which takes both movements to their limiting cases; in some sense invalidating the corner stones of their existence. I will call this other concurrence of participation and computation, 'responsive architecture'; to be vigourously distinguished from flexible architecture, manipulative architecture, or (even) adaptable architecture. I will take my case to some extremes in order to dramatise some of the inherent contradictions, potential banalities, and possible dangers of what remains to be researched and may appear to some as science fiction. My arguments will assume that the reader is familiar with the notions of an 'architecture machine' (Negroponte, 1970), and is acquainted with subsequent experiments (Negroponte and Groisser, 1970). My propositions will include concepts for which we have no historical material or previous experiences from which we can draw perceptions for comparison. Aspects of living in an architecture machine will subsume giving the physical environment things it has never had before: knowledge, common sense, intelligence and any attribute necessary to make the built environment as responsive as a good friend or surrogate self.

Participation and computation have a commonality that is often not dramatized in either case. That is, they both are involved with methods of designing that in some sense shortcircuit or replace the services of a professional architect. In one instance the resident (in the case of a home) becomes an architect for his own needs and, in the other, a machine is emulating the task of a human designer. Both are altering the more traditional chain of events by modifying the first act of a three-part procedure: user-architect, architect-builder, builder-environment. When the user becomes the architect (computer-aided participatory architecture) and the machine can design we are reduced immediately to two stages: user-builder, builder-environment. Whereas, in our case – the responsive architecture – we are insinuating a single scene between user and environment where the building process as well as the design process, as we know them, disappear and are replaced by an architecture characterised by the ability to self-reproduce, to learn, to want, and to play.

Consider two existing (seemingly unconnected) situations:

1) Imagine yourself returning home from work, entering your home and asking your wife (husband, mistress, lover, mother, or whomever) to take the whatchmacallit and put it you-know-where. She (he) most probably knows exactly what you mean for the simple reason that she (he) has an extremely good predictive model of you and can manipulate this model of the context at hand (as described by anything from a frown on your face to the weather outside). A stranger, note, would be at a complete loss and would require much more complete information to handle the whatchmacallit.

2) Ask yourself for examples of everyday physical environments that exhibit some sort of responsive behaviour as the result of inbuilt abilities to compute (as opposed to being under the control of a human operator). Pause. Most probably your first thoughts have included thermostats, but immediately have been disqualified as too simple, binary, and limited in context (he can't mean that). Your next example might include elevators and this is indeed a fascinating example, as some elevators have computing powers that display an uncanny cleverness to handle a large number of inputs. There exist elevators that can interrelate the weight of the car, the time of day, and the number of remaining calls in order to decide, for example, to pass by callers for the purpose of maintaining a twenty-seven minute waiting time in the lobby or the comfort of those in the cab. But beyond elevators? Nothing. Aside from what one might find in industrial environments, hospitals, or airports, elevators remain the only authentic example of computing intimately nested within architecture.

Let me consider the context of the home and extrapolate the elevator example to the extreme condition of where the home is as responsive as the wife. We immediately observe three distinct domains of computation, each of which can be treated reasonably separately and with different levels of elegance: recognition, response, learning. Let us look at aspects of each in terms of what it might be like to live in an architecture machine where interaction is represented as both the supportive (reinforcing) and antagonistic (challenging) behaviours exhibited by both the human and the house.

RECOGNITION

Recognition has dominated a great deal of the study of artificial intelligence. In our case, the recognition of an inhabitant's needs and desires as expressed in a string of spoken words, a smile, a gesture, or a burp, are conspicuous necessities for responsive architecture that, in each case, tax the present frontiers of research into robotics. A more modest, indigenous (to our problem) and exemplary act of recognition is simply "who is here?" The problem of recognising who is present is a fascinating example because it is characteristic of many recognition problems in that it can be attempted in three dramatically different fashions. It can be handled artificially, statistically, or heuristically.

Let us immediately disqualify the first. It circumvents the problem by, for example, embedding an active or passive encoding device in our bodies such that an ubiquitous decoding mechanism can know of our presence by simply matching a short string of bits (32 to embody the present population of the world). Besides disqualifying this means on the basis of Orwellian connotations, we must recognise that a) it can work and b) it bears no relation and affords no contribution to the more subtle aspects of recognition that are necessary: moods, for example.

An experiment on the second method, statistical recognition, is presently being conducted on project GREET (Lavin, 1971), a doorway that attempts to recognise who passes through it. Here, parameters like weight, stride, height, profile, are employed to perform the closest match statistically with a previous encounter or to note a complete mismatch. What is particularly important is that no single parameter exists that can provide a failsafe means of recognition. Height will change if you are wearing high heels, weight will alter if you are carrying packages, and your profile is a function of what hat you are wearing (let alone whether you are walking backwards or forwards). The system, however, can be embellished if we provide further knowledge about: a) the physical environment, e.g., if there is only one entrance, it is highly probable that the same person cannot pass through consecutively from the same direction; b) the natural environment, e.g., if it is snowing, if it is cold; c) the habits of those likely to be found, e.g., he rarely wears a coat in the winter, she has packages delivered. This method can be appended with endless variables and descriptors, adorning the system only quantitatively. There will remain, nonetheless, the issue of, "What if it makes a mistake?" Would you want your house to do for someone else what it does for you?

The third method – a heuristic approach – is the least researched, is the most difficult to implement, has many of the same problems as statistical techniques, and is not (by any means) inherently failsafe. It's primary difference, you might argue, is one of attitude towards the problem. Inputs are employed to make judgments upon how to further examine the available data as well as to build evidence that suchandsuch is true or false. In the previous case all data is collected, sorted, weighted (perhaps) and then mapped upon an existing (and ever growing) table of entries. Heuristically, in contrast, you make guesses, build up evidences and often go back into the real world to solicit further high or low resolution information to develop a hypothesis that it is so-and-so. For example, the height of a person passing through suggests that it is a man, which might in turn suggest that we had better go back and look at the pores on his nose or measure the length of his hair. The sequence continues in the spirit of gathering those clues most likely to damage the supposition being developed and mounting evidence (that it is John).

The problem of recognising "who" has many available and some unexplored inputs like: thermography, radar, machine vision, sonar, galvanic skin resistance. But perhaps more important than the ever increasing range of media for recognition is the characteristic nature of the problem, our attitudes towards it and the dangers involved (does this lead to 'house tapping'?) I propose that these issues are important for all aspects of recognition in responsive architecture but need not be firmly articulated prior to the more basic question: what will it (the environment) do in response to you or me or us?

RESPONSE

Warren Brodey (1967) furnished the first (and almost the last) article on the subject; he launched the notion of 'soft architecture' and 'intelligent environments'. He did not however, give at that time a satisfactory answer to how the environment would respond. His variables included colour, temperature, light, sound – those environmental perfumes that one is usually too lazy to change for oneself at the subtle level (and it is unclear how crucial it will be when they subtly change automatically), or that one finds in a once-is-enough lightshow: "If his heart beat accelerates, the room becomes redder (for example); if his breathing deepens the room takes on a richer hue. As the hue intensifies his heart may beat faster in response to the stimulus (the strength of the colour which changes with his feelings). This total personalised environment is capable of producing a profound experience without brain damage." We hope so . . . but, in the richest of all possible environments (such as that proposed by Brodey), like the chair that alters itself to always perfectly fit your body, does this not lead to a terrible complacency?

Subsequent work of Brodey and his colleague, Avery Johnson, has answered some of these questions and, in some sense, has refuted the initial statements of 1967. Their more recent work (Brodey and Johnson, 1970) emphasises self-referent behaviour (on the part of the environment) and de-emphasises sensing (unless you are measuring 'large muscle' behaviours). The important change in their work, however, is the insistence upon playfulness, noticing a serious lack of it in our present environments. They suggest that environments ought to play with themselves, and present a strong case for a very physical involvement, but they still have not answered the question "What will environments do?" with a wide enough range of examples.

So, then, what will environments do? While I do not propose to be making more than wild guesses, I alert you to the natural tendency towards banal responses and complacent results. Environmental changes that result from aspects of recognition can be categorised within three domains of response: the environmental, the operational and the informational. Each involves interaction between environment and a non-passive user; each must be capable of sampling and contributing to an evolving model of the user(s).

Environmental responses are the Brodey/Johnson kind, the pitfalls of which I have already enumerated. They inherently demand new construction technologies. After all, how can an environment genuinely change as a result of a computation? One of the few examples I know that provides inklings of substantive responsiveness (beyond light, temperature, etc.) in a very simple manner is a small pneumatic shelter developed for an exhibition by Sean Wellesley-Miller. Four photocells were implanted in the doorway to count the number of people coming in and going out of the exhibit. The total number of people present was used as an input to control the air pressure of the structure and accordingly have it shrink or expand as a function of the number of visitors present. Hardly complex, hardly evolutionary, but the notion hints at a responsiveness that can be considered truly 'architectural' and that can be envisaged in a more complex system of pneumatic living.

The operational response is the most practical and could support the platform from which any American politician could gain all female votes. A house that cleans itself or (as John McCarthy (1966) suggested) windows that close themselves when it rains would be practical gadgetry, most probably implementable in a system that does not know me. But a door that serves a martini, or a bed that remains unmade in recognition of my probably returning to it, begins to demand some knowledge of me. While this might sound like the gimmickry of Disneyland or the manoeuvres of a space craft, there do exist simpler tasks of cooking, cleaning and general housekeeping that can be easily handled by simple machines that must be interconnected for the purpose of avoiding their going on what Brodey calls a 'rampage' – turning each other on and off.

The third kind of response that we can expect the home of the future to have is the informational. It has very little to do with what we presently consider the architectural details of living because it does not have any physical embodiment. Examples can be developed through minor projections beyond present communications media and devices: cable television, a clever alarm clock, a good answering service. Or, you can ponder the science fictions of living within an intelligent computer system that can 1) answer questions; 2) synopsise the news within the scope of your interests, the context of a conversation, or your mood at the moment; 3) answer the phone and tell a white lie if necessary; 4) remind you of . . .; 5) suggest good skiing locations; 6) play games; 7) read stories; etc. All of this oozing out of the walls!

Each of the three classes of response has rudimentary, isolated examples that permit us to envisage some of the results of a rapid growth of environmental controls and gadgetry. However, we have no single example of habitation among many integrated responses. We have no idea what an active environment would really be like to live in. Will it leave us alone? Will we be informationally overburdened? Even Appollos, LEMs and the like are of little use in our hypothesising, as they are too goal oriented and do not have to deal with the ambiguities and ambivalence of day-to-day living. What will it really be like to inhabit a physical environment that might be described with such adjectives as: alert, friendly, playful, grumpy, or, simply, 'intelligent'?

LEARNING

Even without a way to envisage authentically responsive behaviour in an intelligent environment, we can be assured that, like the understanding wife (or even the lovable pet dog), it will have to learn, to make errors, and, on occasion, to exhibit hysterical behaviour. Unfortunately, we know very little about how people learn, less about how machines might do it, and consequently I propose to limit the notion of learning (in the context of intelligent environments) to modelling the inhabitants.

Gordon Pask has proposed that we must deal with three levels of model. At the simplest level we are dealing with the house's model of the inhabitant(s) and his (her, their) more frequent visitors. Whether this model is employed to anticipate events, to fill in missing information, or to handle implicit remarks, it can be considered a predictive model, wherein its success and failure can be easily measured as a function of the closeness of fit between the anticipated event (or meaning) and the actual or intended event (or meaning). And, if the number of channels (sensors) into this model is large enough, subtle enough, and redundant enough, the model should be able to cope with context-bound predictions (which out of context might seem contradictory).

In no sense would (or should) such a model be fail-safe or infallible. That is the wrong attitude towards the problem. It is in fact through error-making that the model improves and, at the same time, it must be able to accommodate (even exhibit) whimsical behaviour (that 'information scientists' would call noise). Also, consider that people are not particularly good at modelling each other; it takes a great deal of time (and then works very badly out of context such as in a new culture or with a new age group). A good butler can easily remember how many lumps of sugar you take in your coffee but might be very poor at recognising when a particular mood of yours places privacy before good service.

The next, more complicated level of modelling is the house's model of your model of it. This model is particularly important in the case of inference making, because one tends to leave implicit only those issues which one assumes that the other party understands. In effect, this model grows out of a prosperity of matches between the inferred (by the house) information and the intended (by the human) information. By recognising that the whatchmacallit was an umbrella, preceded and followed by an understood ceremony of events, the house's model of your model of it is reinforced as a result of being able to handle the missing information.

The last level of modelling may appear overly circuitous (it is usually ignored in the study of man-machine interactions). That is: the house's model of your model of its model of you (I will leave it to Gordon Pask to prove to you why we stop at this level). I propose that this level is particularly important for intelligent environments because it is the match and mismatch between this model and level one (its model of you) that can initiate learning. If a window thinks you want it shut and at the same time it can recognise that you think it thinks you want it shut (whether it is a case of your prompting a misunderstanding or simply of your being conscious of a correct supposition) a reinforcement or alteration of this model *must* follow.

How do you construct any of these models? It is one thing to assume (an assumption I am willing to make) a host of sensing and effecting devices that can recognise and respond in a manner that has until now distinguished man from other animals (let alone machines). It is another matter to assume an adaptive and resilient modelling ability (perhaps because our technologies are closer to achieving the latter). We, The Architecture Machine Group at MIT, have struggled with three kinds of model and have learned that two are rather useless and the third, while probably correct, is at present hopelessly unmanageable. I will refer to them as; the determinate model, the probabilistic model, the evolutionary model.

The determinate model is the easiest. In the case of modelling any complex system, a city, a person, or a set of political decisions, one can achieve astounding results through brute force and sheer complication. As in the models of Forrester (1961, 1969, 1971), be they industries, cities, or worlds, the overall model is attained by the coupling of many smaller, manageable interrelationships (models) that we, as humans, can understand as reasonably discrete (even contextless) events. The resulting model is a summation, so to speak, of all these minimodels – a summation that we indeed could not envisage without the aid of a computer. (It is important to note that in this kind of modelling the notion of 'context' is discarded in favour of: "if we knew *all* the interrelationships we would have context", which completely ignores the fact that context is a function of individually ascribed meanings that change from culture to culture, from person to person, from city to city.) Such a model is always at the mercy of its human designer(s), because when it fails it is simply repaired by the addition or subtraction of the parameters deemed necessary. Such dependence and determinism may be appropriate or, at least, revealing in modelling group behaviours, but it is not very helpful in modelling individual people. This is especially true when you are seeking (as we are) autogenic behaviour, inasmuch as we are expecting the machine (the home) to take a corrective course of action without our intervention.

The probabilistic model, meanwhile, circumvents determinism and can exhibit extraordinarily convincing results – its very problem. One can obtain very rapid returns on one's computer programming efforts, when employing probabilities to determine a response and when using inputs to alter those probabilities. While such a model is remarkably responsive in some sense, *it is in no sense learning.* An overly advertised example of such modelling technique is SEEK (probably the closest example I have of living in an architecture machine). In the case of SEEK, the physical environment was composed of some five hundred blocks (two inch cubes) and the inhabitants were gerbils. A simple, computer-controlled prosthetic device could arrange the blocks in three dimensions on a given site (five feet by seven feet). Each location of the site had a probability (for recieving or not receiving a block) such that initially, for example, a block could be placed almost anywhere. As time passed and as a result of gerbil-initiated dislocations of blocks, the probabilities would be altered to reflect such things as: desired circulation routes, open spaces, and where the gerbils spent a great deal of time. The model never settled (in the sense of arriving at an optimum arrangement of blocks that reflected the 'essence' of gerbil behaviour – this non-settling is good), but was instead responsive in a very real sense to the way the gerbils lived within their (unnatural) habitat. Note: no learning was involved.

Learning or evolution (I believe the words are interchangeable) can only really be accommodated in the last kind of modelling, which I have chosen to call 'evolutionary'. I have no example to offer you (those who profess to have examples are deluding their readers). I propose that this latter model is what artificial intelligence is all about. There exist three major barriers (which I believe can be studied separately) that confront us whenever we are dealing with this kind of modelling, whether for robots or intelligent environments.

Problem 1: the sensors and effectors. A machine must have access to the real world through as many channels as possible (at least as many as we have) in order to be able to ascribe meaning to things or events. This is true because we are only able to give meaning to things through *our own* experiences, which we have through *our* senses. And learning cannot take place without meaning. Notice, however, that Problem 1 is something we can work on now, giving machines eyes, ears, a sense of balance, etc.; in some sense biding (or stalling for) time.

Problem 2: learning how to learn. Only recently has the notion of learning how to learn become important in the context of human learning. As most of us were taught 'subjects' we tend not to think of 'thinking' (or learning) as being something one can learn (and teach) in general. The most recent work of Seymour Papert tackles the problem of learning how to learn through computation by making available to children (seven, eight, nine years old) sophisticated computer programming techniques coupled with a wide variety of terminals and toys that the children can program to exhibit some behaviour (more meaningful to the child than text). An exemplary strategy in this work of teaching children thinking is the concept of a 'bug' in a computer program and the subsequent debugging necessary. Learning to debug, as the children do, can be thought of as a form of learning how to learn. I would expect my responsive environment to be able to debug itself, its own model of me and perhaps help me to debug mine of it.

Problem 3: wanting to learn. Whether it is for reasons of survival or because of an intrinsic desire to play, humans apparently want to learn, which in turn fosters, if not contains, some of the ingredients to do it. We do not know how to make machines want anything. It might not be a good idea to find out. I am not sure I would be pleased if my house *wanted* company. I am sure I would like it to *want* to be responsive, I do not know how to do it.

CHATTING WITH COMPUTERS

Christopher Evans

I am a psychologist, and I work sometimes with human beings and sometimes with computers, and sometimes with both; my principal interest is finding better ways of effecting communication between people and computers. By communication I do not mean just idle conversation or exchanging very simple information, but I mean communication in the form of anything from the interactive use of terminals for mathematical problems, to using terminals in design problems, and computers as methods of simulating human personality, etc.

Good working computers have been in use for a quarter of a century: it is 25 years since the first major systems were developed which would do a useful job of work. At the time when they were first developed they were called 'electronic brains', and people were tremendously impressed with their potential. There was a wave of speculative material and newspaper articles of one kind or another which suggested that we were on the threshold of a transformation of human society because of these developments of 'electronic brains'. But now, 25 years after, what has happened to this revolution? Why is it that computers, after a quarter of a century of being real things, have made so little impact on ordinary people?

In fact, they have made little impact not just on ordinary people but on most people in any kind of life. One can take that a step further and say that they have had relatively little impact even on most scientists. If one talks to the majority of scientists today, one finds that their usage of computers is minimal and largely unimaginative. Now this is curious because the computers *have* worked and have been capable of doing many jobs better and quicker than human beings. So why is it that there has been this considerable lag in computers having any impact on society?

I think there are three hurdles that have stood in the way of widespread computer applications: cost, size and language. Until you can afford the things, find a place to put them and then learn to use them, you are stuck. Cost – for a really good computer which would do anything useful you might have to spend up to £250,000. It is not as much as that now, but that's a typical figure. Size – again until quite recently the things have been too big; universities, laboratories, hospitals, schools, businesses, anyone who might want to use a computer – they have all got crushing space problems. Language – this is really an awful objection because it is tied up in the whole question of training people how to use a computer, and there is no getting away from it. One gets the impression that before one can make really good use of the computer – or even use it at all – one has got to learn a special language; one has got to have some kind of special mathematical ability, some marvellous gift, or have been given a complicated course in programming. There is no doubt that this is partly the reason for the doubt surrounding computers, because people feel: "How can I possibly find the time to learn about computers – and even if I can find the time, am I *capable* of learning how to use them?" These are real objections and they account largely for the very slow impact of computers in non-specialist scientific circles.

But some very interesting things have happened in the last few years which have changed the whole face of computer technology and computer theory. For instance, the principle of computer time-sharing has been developed. One original problem was that if one wanted to use a computer, one had to come from some distance to the place where the thing was located, and then to compete with other users on the system. Now there could be a way round that, of course. For example, if we could manage to arrange for the computer to divide its time up effectively amongst a lot of different users, then we could have a number of access points on the outside of the computer, and get a lot of people working on it.

Now even that means the individual has got to go to the computer, so another way out of this is to run a line down and have the access point, a terminal, some distance away, simply linking it up by a wire. There can be as many terminals as your system will support. This would work for a while, but in due course we are going to get real problems because of the length of wire that is required. This still means that there is a limitation – one has to get reasonably near to the computer before one can use it properly.

Fortunately, however, the Post Office has been kindly laying wires down all over the country for many years, and if one has the right kind of equipment – which will convert the signals from a terminal into the signals that will drop down telephone lines – one can use the Post Office network and therefore get access to the system anywhere there are telephones. Now we have got rid of the cost and size objections immediately because, in this capitalist society of ours, one finds that there are business organisations which will invest money in computers and set them up and then rent them out to people who want to use them. So someone else buys the computer and finds the place to put it.

We were still troubled with the language barrier of course, but as soon as people realised that this kind of multi-access computer was going to be possible, a lot of thought immediately went into developing languages which would allow people to program and to use computers with the minimum of training, and also in a language form as close as possible to that of normal human communication. Thus a lot of effort has been put into developing computer languages which approximate to English. This has made an enormous difference. I do not think it has made a great deal of difference in circles that were fully computerised, that were wrapped up in computer technology anyway, but it *has* made a tremendous difference to people and organisations that could never get hold of computers before.

I have had to make this long introduction because it makes no sense to talk about the kind of work which I have been doing unless one can see how easy it is to do nowadays. If one has, in effect, a portable computer it is possible to go into any place where it may never have been possible to use a computer before, and conduct experiments. One can see how computers can be used in the wilds, or in environments where they have never before been accessible. Some of the places I am talking about are places like hospitals, schools, business organisations, and so on, The area that I particularly want to discuss here is the use of computer terminals, and these new types of systems, in hospitals.

The question I have been pursuing in my work is, "What is the best method of communicating with the computer – what is the best method of getting the information into the computer, and of the computer putting it out?" It depends, of course, very much on who you are; what type of person you are. If you are thoroughly *au fait* with computers, then you want the information to come out as quickly as you possibly can, and you probably want it in coded form so that you do not have a great deal of verbiage. In this way you can get precisely to what the computer is trying to say to you, and you can get your ideas in very simply and very quickly. That approach, however, is of interest only to a very limited and very small section of society, and the very much larger section of society who can now get access to a computer – in principle anyway – want something different, or at least one suspects they want something different.

At the Department of Medicine in Relation to Mathematics and Computing at Glasgow University, for example, a lot of thought has gone into the question of whether or not patients in a hospital could be interrogated about their illnesses by computers. We are talking now about real, ordinary patients, who drift into a hospital expecting to see a doctor. Would it be possible for people to have questions put to them by computer – either by voice or by teletype or something like that – and would they be prepared to accept this? This seemed to me to be a very interesting problem because it raised a whole lot of questions about ordinary people's reactions to computers, and I thought that the hospital setting, where people might be most anxious and least tolerant of peculiarities of the computer kind, might be a very good place to jump in.

So it was arranged to put a teletype terminal into a hospital, and we set about writing the program for it. The first thing to do, was to take the problem itself – which is a diagnostic problem in this case of whether or not patients had a duodenal ulcer. It seems that people coming to hospital with suspected ulcers is a very common condition. It also so happens that a diagnostic problem (or the first stage of a diagnostic problem) is not difficult to specify, and it should be quite easy to determine what are the steps in the diagnostic process.

We thought that the first thing to do was to try to find out what questions a doctor would ask a patient, but, as we discovered, getting a description from people as to what they do almost automatically, is very much more difficult than one might imagine. Even in the case where what you are asking them to do is to tell you what they say in a particular context, it is very difficult for them to firmly describe the process. I suppose its rather like trying to describe how you ride a bicycle, or something like that; once you start to describe it you find yourself in trouble.

Now the only thing to do was to sit in on a number of diagnostic sessions, and just take a notebook and write down the questions. When I did this, the doctor conducted the interview off the top of his head, without trying to think about it, and then I found that there was a very nice, simple picture building up. We could identify quite clearly the steps in the tracing of the history. We also found that they were fairly simple steps, and there was not a great deal of disagreement between one physician and the other. Once we had got that clear, then all we had to do was to write a program putting these questions, and make it a branching program such that if one asks "Do you have a pain in the stomach?" and the answer is "yes", one goes on to another particular question, and if the answer is "no", then one goes on to something else. That side of it did not turn out to be difficult to do, but the next problem was how we should phrase these questions. What should one do – should one get the information over as quickly as possible, or proceed very slowly?

There was a certain constraint in that the patients we were going to deal with were fairly simple people, who came into the Out Patients' Department of a perfectly ordinary Clydeside hospital, which people who work on the Clyde in ship building come into when they have been referred there by their doctor. Now the national figure for people who can use a typewriter is probably something like 5–10% of the population, and it was fairly clear to us that the average patient walking into the hospital in Glasgow and sitting down to a teletype, wasn't going to be able to use it just like that.

I should say that we thought it important that the whole thing from start to finish should be conducted by the computer – the patient should come in, and the computer should introduce itself and should go right through the whole sequence and say "Cheerio" to the patient at the end, and out they would go, with no doctors or technicians standing there or holding their hands or anything like that. The justification for this view is that the experiments go beyond the field of medicine. We are interested not only in medicine but also in psychiatry, in teaching, or in any situation where people who had never been near a computer before in their lives might suddenly find themselves faced with one. The question is; how would they feel about this? How easily would they use the equipment? How much information would the computer have to give about itself? I thought that the most sensible approach was that one should make the thing as human as was reasonably possible, but never make it pretend to be anything other than a computer. In other words, it could assume a form of human dialogue, but it would say "I am a computer".

Because of the difficulty about typing abilities we have modified the face of the teletype so that there is a simple mask over it with a small number of buttons showing. We have three buttons in this case, with 'yes', 'no' and 'don't know' on them. We have reduced the teletype to single button operation – there is no necessity to operate carriage returns or any of the other things which people consistently forget about. The mask is a very simple thing, costing about £2, and it converts the teletype into a simple, interactive, one-button system.

With the teletype, of course, the computer material does not come up in a flash. This is a very important shift from practically everything that has been done with computers in medicine up to now, where the researchers made the mistake of thinking that people must have an ultra-rapid response from the system. The quickest way to do that is, of course, to have the question printed out, perhaps on a slide, and at the appropriate moment the slide comes up on a screen with a mass of text. In practise this does not work very well because, firstly, people don't always read it, and, secondly, they tend to get panicked by it.

So our material is generated sequentially by the teletype typing it out, and it begins (Figure 1), "Hello". Why shouldn't a computer say "hello", its a perfectly good way of opening communication? It goes on: "This is an experiment to see whether computers can help doctors diagnose illnesses. We'd very much like you to help us with this". Now this is very cunning, because people usually are pretty nice. If you ask them to help you, they will. So we say, "We would very much like you to help us with this", and people always say "yes". It continues, "If you are prepared to, would you push the button marked 'yes', but if you'd rather not, push the button marked 'no' ". This has been generated at ten characters a second, which is slow for well educated people to read, but it is about the right reading speed for the majority of people.

We then say again, "But we would like you to help us. If you don't understand what to do press the 'query' button. Go ahead and push one of the three buttons". If they now push the query button, the thing explains in a sort of 'Noddy in Toyland' language over again, and it goes into great detail about what it is all about. If they say "yes", the computer says, "Good, thanks very much," and thanks them for doing something. If they say "no", it says "Well can't I persuade you to change your mind, because we'd like you to do this?"

If the patient agrees to continue, it says, "I hope you'll find it interesting, we'll find it very helpful to us". There is a very important step that it now does. It says, "This is a computer talking to you". It states its credentials from the beginning and explains its limitations: "I can ask you questions but I can only understand simple answers. You can only talk to me by pushing the 'yes', 'no' or '?' buttons". In other words, it says, "What I can do is so-and-so, and what I can't do is so-and-so," and people find themselves clicked into the picture quite easily with that. Now

```
HELLO.  THIS IS AN EXPERIMENT TO SEE WHETHER COMPUTERS CAN
HELP DOCTORS TO DIAGNOSE ILLNESSES.  WE WOULD VERY
MUCH LIKE YOU TO HELP US WITH THIS.  IF YOU ARE PREPARED
TO WOULD YOU PUSH THE BUTTON MARKED 'YES', BUT IF YOU
WOULD RATHER NOT, PUSH THE BUTTON MARKED 'NO', BUT WE WOULD
LIKE YOU TO HELP US.  IF YOU DON'T UNDERSTAND WHAT TO DO PRESS
THE '?' BUTTON.  GO AHEAD NOW AND PUSH ONE OF THE THREE BUTTONS?
(yes)

GOOD. THANKS VERY MUCH. I HOPE YOU'LL FIND IT INTERESTING
WE'LL FIND IT VERY HELPFUL TO US.

THIS IS A COMPUTER TALKING TO YOU.  I CAN ASK YOU QUESTIONS
BUT I CAN ONLY UNDERSTAND SIMPLE ANSWERS, SO YOU CAN ONLY
TALK TO ME BY PUSHING EITHER THE 'YES', 'NO' OR '?' BUTTON.
YOU SHOULD ONLY PRESS THE '?' BUTTON IF YOU DON'T UNDERSTAND
THE QUESTION I AM ASKING YOU.  NOW, DO YOU UNDERSTAND
EVERYTHING I'VE SAID UP TO NOW?  PUSH ONE OF THE BUTTONS?
(yes)

THANKS. NOW I WANT TO ASK YOU ONE OR TWO QUESTIONS ABOUT WHY
YOU ARE HERE.  PLEASE REMEMBER THAT THIS IS ALL IN COMPLETE
CONFIDENCE JUST AS WITH DOCTORS.  ONLY DOCTORS AND NURSES
WILL SEE YOUR ANSWERS.  ARE YOU QUITE HAPPY ABOUT THIS? PUSH
ONE OF THE BUTTONS?
(yes)

GOOD.  BY THE WAY, DON'T FEEL THAT YOU HAVE TO PRESS
THE BUTTON THE VERY INSTANT THAT I HAVE FINISHED TYPING.
IF YOU FEEL YOU NEED TO THINK A BIT BEFORE YOU ANSWER, OR
NEED TO READ THE QUESTION OVER AGAIN THEN GO AHEAD AND DO SO.
TAKE YOUR TIME IF YOU WANT TO.  REMEMBER, HOWEVER, THAT
I CAN'T TALK TO YOU AGAIN UNTIL YOU HAVE PUSHED ONE OF
THE BUTTONS.

WELL, FIRST OF ALL THEN, IS ONE OF THE REASONS YOU VISITED
YOUR DOCTOR BECAUSE YOU WERE SUFFERING FROM DISCOMFORT OR
PAINS IN THE STOMACH?
(yes)

COULD YOU TELL ME, DOES THE PAIN OR DISCOMFORT TEND TO COME
AND GO - IN OTHER WORDS ARE THERE SOME TIMES, DAYS OR WEEKS
WHEN YOU'VE BEEN COMPLETELY FREE OF THE PAIN?
(no)

SO YOU HAVE NEVER HAD, SINCE THE ILLNESS BEGAN, TIMES,
SUCH AS DAYS OR WEEKS, WHEN YOU'VE BEEN COMPLETELY FREE
OF PAIN?  PUSH 'YES' IF YOU HAVE HAD TIMES FREE OF PAIN
AND 'NO' IF YOU HAVE NOT?
(no)

I SEE.  TELL ME IS THE DISCOMFORT RELATED TO TAKING FOOD - DOES
EATING HAVE SOMETHING TO DO WITH IT?  PUSH 'YES' OR 'NO'?
(yes)

WHEN YOU HAVE THE PAIN OR DISCOMFORT DO YOU GET
RELIEF FROM SUCH THINGS AS INDIGESTION TABLETS, LIQUID
MEDICINES OR BAKING SODAS?
(no)

TELL ME DOES THE PAIN EVER WAKE YOU IN THE EARLY HOURS OF
THE MORNING?  PLEASE PUSH 'YES' OR 'NO'?
(no)

I EXPECT YOU KNOW WHERE THE NAVEL IS, BUT I WANT TO BE QUITE
SURE ABOUT THIS.  PUSH 'YES' IF YOU KNOW WHERE THE NAVEL IS
AND 'NO' IS YOU DON'T KNOW?
(no)

WELL, SOME PEOPLE CALL IT THE BELLY-BUTTON.  NOW DO YOU KNOW
WHERE IT IS?
(no)

I'M SORRY IF I AM NOT MAKING MYSELF CLEAR. PERHAPS IT
WOULD BE BETTER IF WE STOPPED THE EXPERIMENT HERE.  DON'T
WORRY ABOUT THIS FOR YOU HAVE BEEN VERY HELPFUL IN TAKING
PART TO THIS STAGE.  COULD YOU TELL THE NURSE OUTSIDE THAT
YOU HAVE FINISHED.  GOODBYE AND THANKS AGAIN.
```

Figure 1 An example of computer-patient conversation. The patient's responses are shown (lower case) as written replies, but in practice the patient merely presses a coded button. The example demonstrates the computer's escape routine for a patient who insists that he does not know where his belly-button is.

they push a button. If they don't understand, the thing explains and then says, "Now I want to ask you one or two questions about why you are here".

It then goes into something I thought was very important, and we've since learned in fact is important – the business of confidence. People have got a paranoid feeling about computers, quite reasonably perhaps, and one might expect a patient in hospital could feel particularly paranoid about this, so the computer says, "This is all in complete confidence; do feel quite happy about that". Then it says, "Take your time, don't feel you have to rush in your answers as soon as I've finished typing. Think about the question again if you like", and so on.

At all these points people have got get-outs. I thought it would be unfair, perhaps unethical, to imprison people in a room with a terminal and force them to use the thing. So I thought that if they really showed signs of being unhappy – either by giving a peculiar sequence of responses, or by saying more than once that they didn't want to take part in this – they could get out. If it is clear the patient doesn't understand what is happening and is just punching away at buttons, desperately anxious to get out, then the thing says, "Well, that's alright. Thank you for what you have done, now get up and go and see the nurse outside", or something like that.

If the patient seems happy, the computer now gets down, after that long introduction (which incidentally allows the person to acclimatise to the system and get used to pushing the buttons and settle down), to the real medical business of getting the information. Depending on the patient's answers, so the program moves along particular trees of enquiries.

There are some difficulties sometimes, because one can't be sure what it is that people understand. At one point the program says, "I expect you know where your navel is" (we are trying to identify the source of the pain here), "but I want to be quite sure about this. Push 'yes' if you know where the navel is and 'no' if you don't". Now I was certain there would be some people who wouldn't know where their navel was – people being what they are – and that an even bigger percentage would push 'no' to a stupid question like that. So to those who said 'no', it responds, "Well some people call it the belly-button. Now do you know where it is?" In fact we did find that some of the patients did not know, or said they did not know, where the navel is, and were quite happy to be told the explanation of it. If the patient insists that he does not know where the belly-button is, then the program gives up, thanks him for having taken part thus far, and sends him off with a polite "Goodbye". Perhaps I should just add that, so far, we are using the computer only to collect information, which is then passed to a doctor for a decision.

When we installed the computer terminal in the hospital for the first time, we stuck to the principle that it should be useable by anyone at all, and so we just asked the first person who happened to be in the waiting room that day to use it. This was an elderly lady, trembling with anxiety – remember she is there because of a suspected ulcer, and people are naturally anxious in the hospital situation anyway – but she agreed to try the experiment. So I took her along to the terminal, made sure she could read the print-out adequately, and then just left her alone with the terminal. I was actually able to watch her unobtrusively, and I saw that, after the first computer query, she paused for a moment and then pressed a button; the computer started up again, she paused, pressed a button again, and so on. She went right through the interview without turning a hair. This was the remarkable thing that we found – patients can be put absolutely 'cold' in front of the terminal, and they will go through a fairly complicated interview lasting about twenty minutes, with no difficulty at all.

Now, being an optimist, I had thought that a large percentage of people would be quite prepared to go through this exercise, just because they wanted to help and because they thought they might be able to get better from it. But it has become very clear to us, in many experiments now, in different situations and different parts of the country, that it is much more satisfactory than that. The interaction between the individual and the system is quite dynamic. We have tested people afterwards, on their reaction to the system, and we found that nobody disliked it. Most people said they were in favour of it, and a number even said they actually preferred the computer to a human doctor.

This favourable response poses some very interesting questions, which are not just to do with medicine, but are to do with computers in general, and people's reactions to quasi-intelligent machines, which will concern us increasingly over the next decade or so. Why should it be that these people, who have come to hospital expecting to see a doctor but are instead put in front of a machine, say that they find the machine preferable to doctors?

I think there are a number of reasons for this preference. For instance, there is a lack of embarrassment when using a machine – in particular because the machine does not judge one in any way; doctors, being human beings, naturally seem to be judging one in an interview situation. Another reason is that one is not being hustled by the machine, whereas one can easily get the uncomfortable feeling of tying up the doctor's valuable time. Perhaps the most important reason for people preferring the machine is that it does not present any class barriers; the doctor is a privileged, educated, well-to-do member of the community, but the vast majority of his patients are on the other end of the social scale, and it is difficult for them to establish an easy communication with him.

We are now going on to experiment with different types of interface. For example, we have found a clear preference for a cathode ray-tube terminal, which puts up the material in just the same way as the teletype, but silently. We are also trying voice output, using pre-recorded tapes.

Computer applications of this type are obviously going to have profound consequences, not only in medicine, but in teaching and many other professions. Although we have a long way to go yet, in the medical context I would not be prepared now to say that the human being is *de facto* the best diagnoser of an illness. And the same argument no doubt applies, or will apply, in numerous facets of professional life.

EXPERIMENTS WITH PARTICIPATION-ORIENTED COMPUTER SYSTEMS

William J. Mitchell

Sometimes we can search for solutions to problems by the direct manipulation and observation of the object, system, or situation which concerns us. But in design and planning, this approach is rarely practicable, and we rely instead upon the use of various types of models. These models are purposefully unrealistic forgeries. They are forgeries in the sense that we construct something to represent something else, and they should be unrealistic since we reduce the complexity of reality to a point where we can deal with it. Thus in a given situation, the model which we adopt depends on those aspects of the whole which interests us, and we eliminate details not relevant to those aspects. A satisfactory model for a given purpose combines sufficient simplification with adequate realism, relevance, and accuracy to enable us to accomplish that purpose.

Facility in constructing, understanding and using certain relevant types of models is one of the necessary skills of a design professional; architects must work with plans, elevations and sections, electrical circuit designers with circuit diagrams, economic planners with systems of equations, and so on. Characteristically, the models used by design and planning specialists are quite difficult for laymen to understand, and this forms a considerable barrier to their effective participation in the design process.

This paper describes some work with computer systems which, we hope, will result in an opening-up of architectural and urban design processes to wider and truer participation by making it possible for non-specialists to comprehend and directly manipulate quite powerful models of the environment.

THE ROLE OF COMPUTERS IN FACILITATING PARTICIPATORY DESIGN

How can computer systems help us to achieve this goal? In order to answer this question it is helpful to consider the diagram of design and planning processes illustrated in Figure 1 (after Newell, 1966). The problem situation which initially exists is represented by the node S(1). Conceivable future situations are represented by the set of nodes S(2), S(3), S(4), . . . S(N). There exists a set of actions A(1), A(2), A(3), . . . A(P) which transform situations in to other situations. Our task is to discover an action or sequence of actions leading to some desired future situation S(G). Then in principle, the range of conceivable future situations, and their relations to each other may be represented by a tree as shown. Design or planning is resolved into a process of exploring at least part of this tree by asking a sequence of questions – IF we take action A(X) THEN which situation S(?) results?

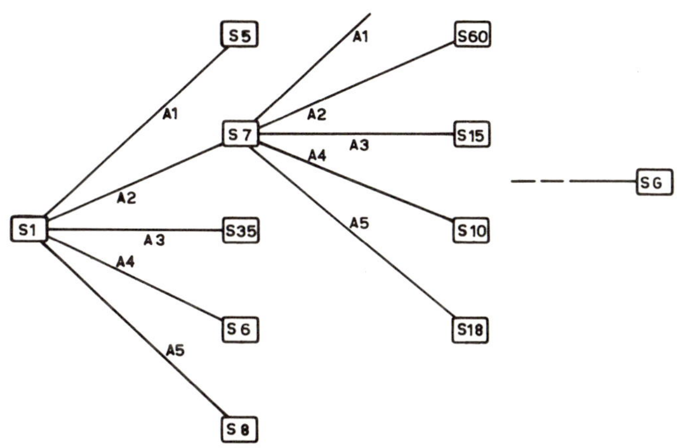

Figure 1 Design or planning as a process of search.

Of course this is quite a gross over-simplification, but, conceptually at least, it does provide us with a clear and straightforward set of criteria for evaluating models and media for the construction and manipulation of models. We simply require that a model should answer our if/then questions with sufficient speed and accuracy, within desired bounds of cost, and in an appropriate format.

Now different modes of design and planning impose different requirements of speed, cost, accuracy, and format. A real-time traffic control system may require answers in seconds, but often it may be possible to allow days, weeks, or months to analyse the consequences of a particular planning policy. The designer of an automobile spends (relative to the price of the product) much more on investigating alternatives than the designer of a house. Participatory

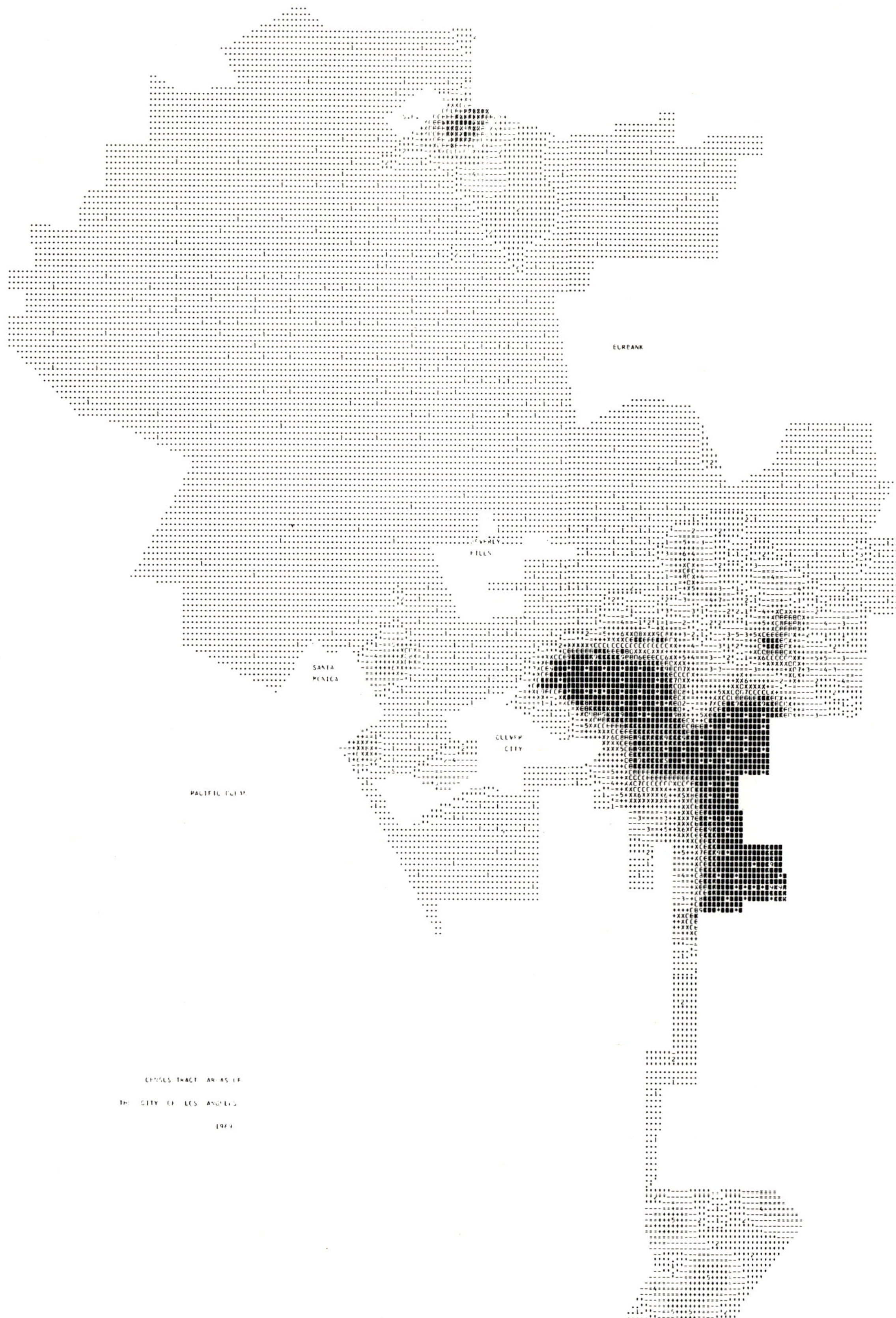

CITY OF LOS ANGELES

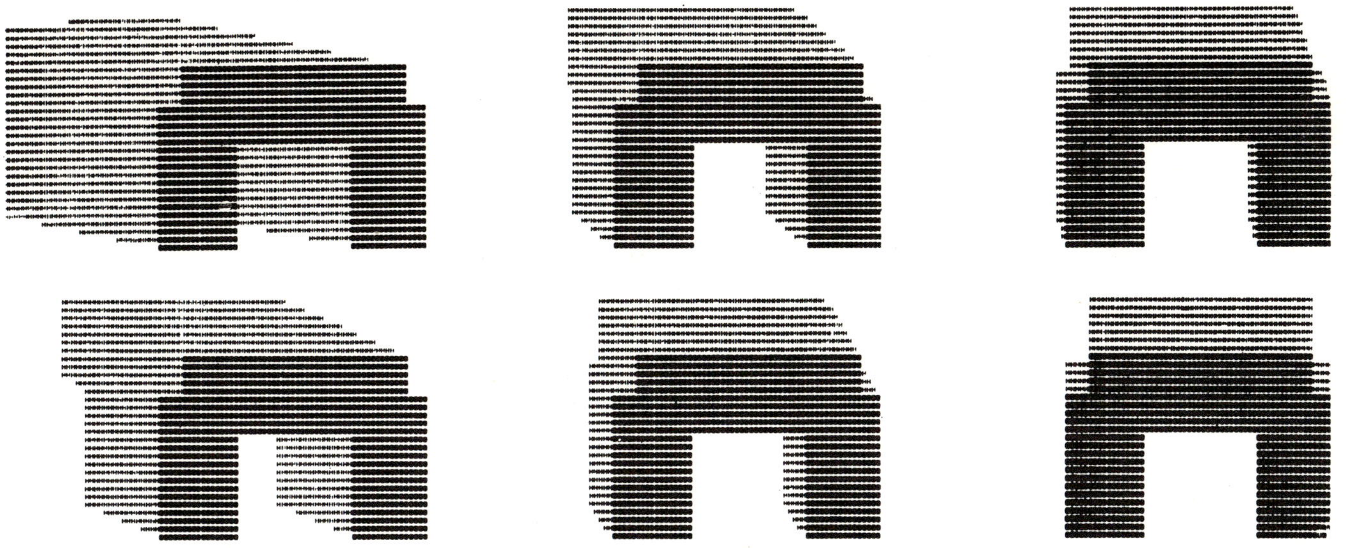

Figure 3 Use of computer system to answer a straightforward question by system simulation – shadow patterns (developed by Murray Milne and Robin Liggett).

design implies particularly stringent speed and format requirements; answers must be available rapidly enough to allow speedy and facile exploration of a wide range of alternatives, and must be presented in formats readily comprehensible to non-specialists in design. Combination of the high speed processing and versatile graphic information display capabilities of modern computers seems then to offer many exciting possibilities for the development of participatory design.

SYSTEMS TO ANSWER IF/THEN QUESTIONS IN DESIGN AND PLANNING

Design situations which we encounter may, for our purposes here, be divided into two classes; those in which particular actions will have clear, and in principle relatively simply predictable, consequences, and those in which actions produce rather more complex outcomes. In other words, our if/then questions may or may not have straightforward answers. We should be able to deal with both cases.

Computer systems have long been used with success for rapidly obtaining straightforward answers, by performing information retrieval, system simulation, or design optimisation tasks. We can ask, and receive almost immediate answers to such questions as, "What is the ethnic make-up of census tract X?", "What shadow patterns will be generated by this arrangement of buildings on a site?", and "What is the cheapest standard rolled-steel joist capable of supporting load X over distance Y?" Figures 2–5 show examples, from projects developed at UCLA's School of Architecture and Urban Planning, of the use of the computer in this role. In each case, the format in which the answer is presented is such that it can be readily comprehended by non-specialists.

Consider now, a rather more complex and difficult problem. Figure 6 illustrates a computer program, developed by William Newman at the University of California at Irvine, to deal with a class of planning problem currently rather important in California politics – the division of a map into zones for purposes of political reapportionment. In the example shown, a map of Los Angeles is displayed on the screen, and the operator, using a Rand tablet, draws upon it the outline of a proposed political district, or a modification of an existing district. On the basis of demographic data and a predictive model of voting behaviour, the system responds by displaying the population of the proposed district, and the percentages expected to vote Republican and Democrat. In this situation, there is a rather complex set of relationships between actions and their predicted consequences. Firstly, we must recognise that conflict of objectives exist; the Democrats would seek one outcome, the Republicans another, some criterion of equity would presumably imply a third, and the Peace and Freedom Party would be most unimpressed by the whole exercise. Secondly, the program must make predictions concerning the behaviour of a system which is considerably less determinate than, for instance, sun motion, and about which we have limited and necessarily incomplete knowledge.

Where design or planning actions have such complex, unclear, and perhaps conflicting outcomes, it becomes

Figure 2 Use of a computer system to answer a straightforward question by information retrieval – percentage minority enrolments in Los Angeles schools.

Figure 4 Use of a computer system to answer a straightforward question by system simulation – Cityscape; an on-line, real-time, user controlled coloured perspective display simulating movement through a city, using General Electric's NASA moon-landing simulator (application developed by Peter Kamnitzer).

meaningless to speak of or seek one "best" solution to a problem. Characteristically, such problems are solved by making trade-offs, negotiating, and weighting the effects of incomplete knowledge and the possibility of unexpected situations arising. Perhaps the most exciting potential of computer systems for participatory design is in such situations, by providing a medium through use of which processes of trade-off and negotiation are facilitated.

Even quite unpretentious computer systems seem to be able to perform this role surprisingly well; Figure 7 shows some typical output from a simple system, CLUMP 3, which I developed to assist in problems of spatial arrangement, and now use as a tool for teaching basic design. It operates through an IBM 2741 typewriter terminal. Users interact with the system by entering statements in English and receiving replies also in the form of English language statements. Basically, the output consists of a set of spaces comprising a building, attached to each member of which is a series of statements defining its "locational attributes". By performing some operations on this data, the system determines and prints out the pattern of grouping of spaces implied by a given set of input statements. The user can experiment with the spatial system he defines by adding and deleting spaces and locational attributes, and observing the effects of these actions on the pattern of grouping which emerges.

The object is not to use the computer to find some "best" arrangement of spaces, but rather to enable the user to carry out a kind of "Socratic dialogue" in order to clarify his understanding of the problem, by exploring the implications of alternative ways of formulating it – i.e. of answering the questions, "What special places do I wish to create in this situation?", and "How should these places relate to each other?". Where conflicting locational objectives exist (as they usually do), use of this system facilitates identification of the precise character of these conflicts, and provides a clear basis for trade-off, negotiation, and compromise. Experience with its use as a means of introducing architecture students to questions of conflict of objectives in design has been most encouraging (Mitchell, 1970a, 1970b).

Figures 8 and 9 illustrate some rather more ambitious systems for answering complex questions, also under development at UCLA, dealing respectively with site layout, and land use and transportation planning problems. In both cases, the user inputs a proposed design by drawing it with a Rand tablet or light pen. Various checking procedures are then executed, and evaluations (in relation to a number of different relevant criteria) are displayed on the screen. These checking procedures, together with associated data bases, embody a great deal of specialised knowledge which would normally only be accessible to professional specialists, but it is now made available to any users of the systems. Again, it is intended that this feedback of "thens" provided by the systems in response to the "ifs" postulated by the users will facilitate processes of trade-off, negotiation, and conflict resolution, leading to convergence on solutions acceptable to all concerned.

COST AND ACCURACY LIMITATIONS

My discussion so far has been largely concerned with illustrating the potentials of computer systems for participatory design due to their capacity to provide answers to our if/then questions at high speed and in appropriate formats. But these capabilities are of little use unless the answers provided are cheap enough and accurate enough for our purposes.

In the past, the cost of computer graphics terminals of sufficient sophistication to be useful in design has been a severe limitation on their use. However, costs seem to be rapidly diminishing, and it is now possible to buy excellent terminals in the 10,000–20,000 dollars range, as compared with 100,000–200,000 dollars necessary in the past. There seems no reason to doubt that this downward trend will continue well into the future.

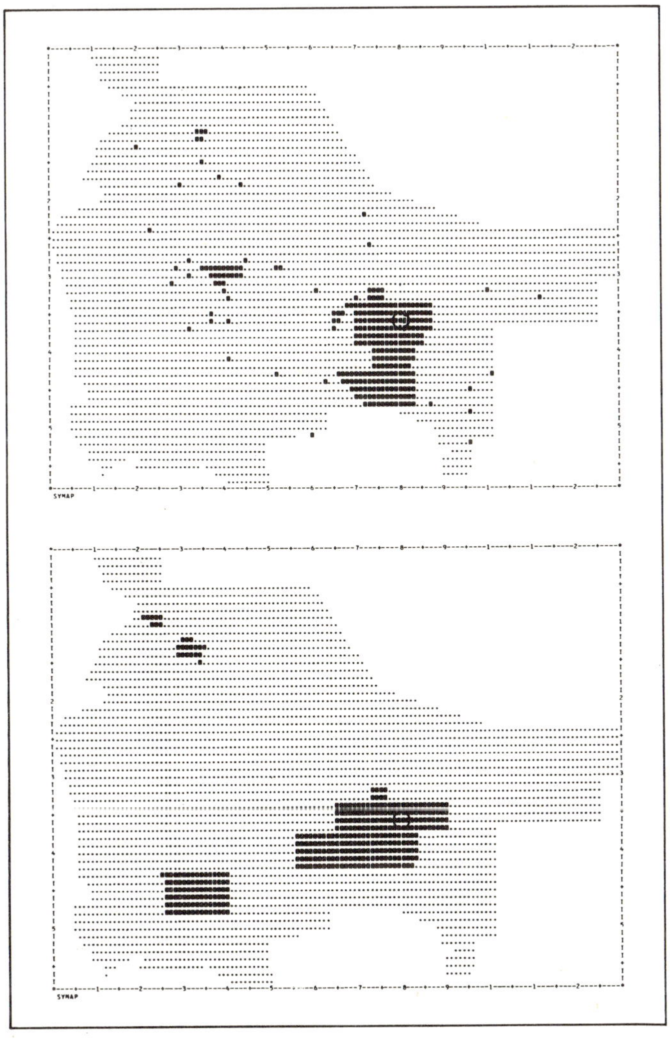

Figure 5 Use of a computer system to answer a straightforward question by system optimization – assignment of pupils to schools in Pasadena, California, in order to produce required integration whilst minimizing busing. Top map shows locations of students at present travelling to a school (marked by circle). Bottom map shows solution derived by an optimal assignment algorithm. (From a study by Robin Liggett.)

Of course no answers provided by a computer system can be better than the models and data bases from which they are derived, so the practicality of computer systems for participatory design depends directly on the state of the art in these areas. Thus in the fields of urban design and planning, we should expect some exciting payoffs to result from the intense activity presently focussing on urban modelling and data base construction.

Figure 6 Use of a computer system to answer a complex question – finding acceptable locations for political district boundaries in Southern California. (Program by William Newman, University of California at Irvine.)

```
PERCENT SIMPLICITY   10.526 PERCENT DATA LOSS    3.226 BENEFIT RATIO    3.263
DO YOU WANT TO SEE A FURTHER LEVEL OF ANALYSIS?
YES OR NO        >yes
ANALYSIS OF CLUSTER LEVEL   2 COMPLETED...

*********************************************************************************
A SUBURBAN HOUSE
CLUSTER LEVEL   2        AT THRESHOLD     88
*********************************************************************************

CLUSTER NUMBER   2. 1

MEMBERS                         LINKS            CONNECTED TO
  3 BREAKFAST/INFORMAL EAT        2* 14, 13,
 13 INFORMAL ENTRY                1*  3,
 14 KITCHEN                       1*  3,

ATTRIBUTE LIST   (ATTRIBUTES POSSESSED BY  100.000 PERCENT OF MEMBERS)

PART OF GROUP LINKED BY FOOD PREPARATION/SERVING/DISPOSAL ACTIVITIES
FACE EAST FOR MORNING SUNLIGHT
PART OF AREA SUPERVISABLE BY MOTHER WHILST WORKING
PART OF ACTIVITY-ORIENTED ZONE

PARAMETERS...IAT   15,ASF    3,APR   80.000,LIN    2,PLN    3,ALN    2,
PCN   66.667,PRD    0.0  ,MOA   3
NEXT PAIRS TO JOIN WILL BE...
( 14, 15) (  3, 20) (  3, 17) (

*********************************************************************************
```

Figure 7 Use of a computer system to answer a complex question – CLUMP3, a system for exploring spatial arrangement problems (developed by William Mitchell).

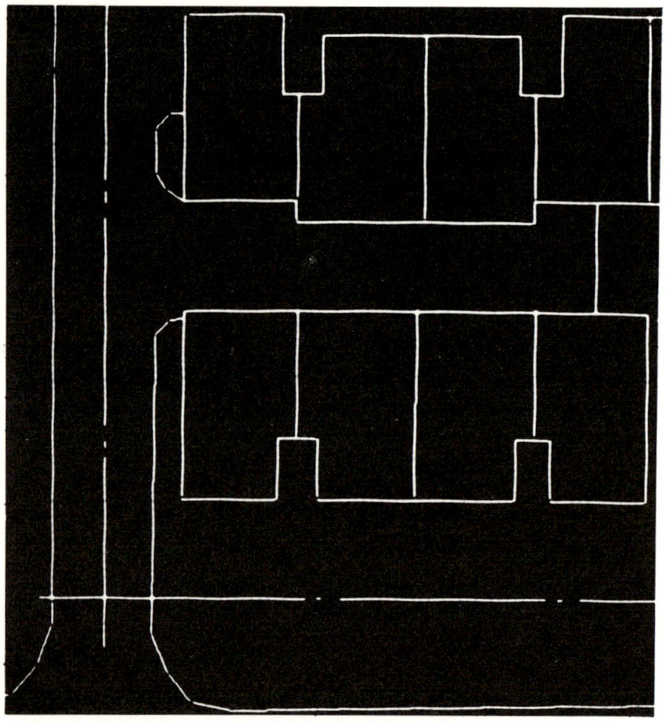

Figure 8 Use of a computer system to answer a complex question – SITE-PLANNER, a system for exploring site-planning problems (under development by William Mitchell).

THE FUTURE

Most of the systems discussed in this paper are still very much in the developmental stage, and some are really little more than demonstrations of possibilities and potentials. There is, as yet, very little real-world application. But the prospects seem encouraging – our models and data bases are getting better, the cost of interactive computer graphics is diminishing, and we are gaining experience and expertise constructing participation-oriented systems. I think we can expect rapid progress.

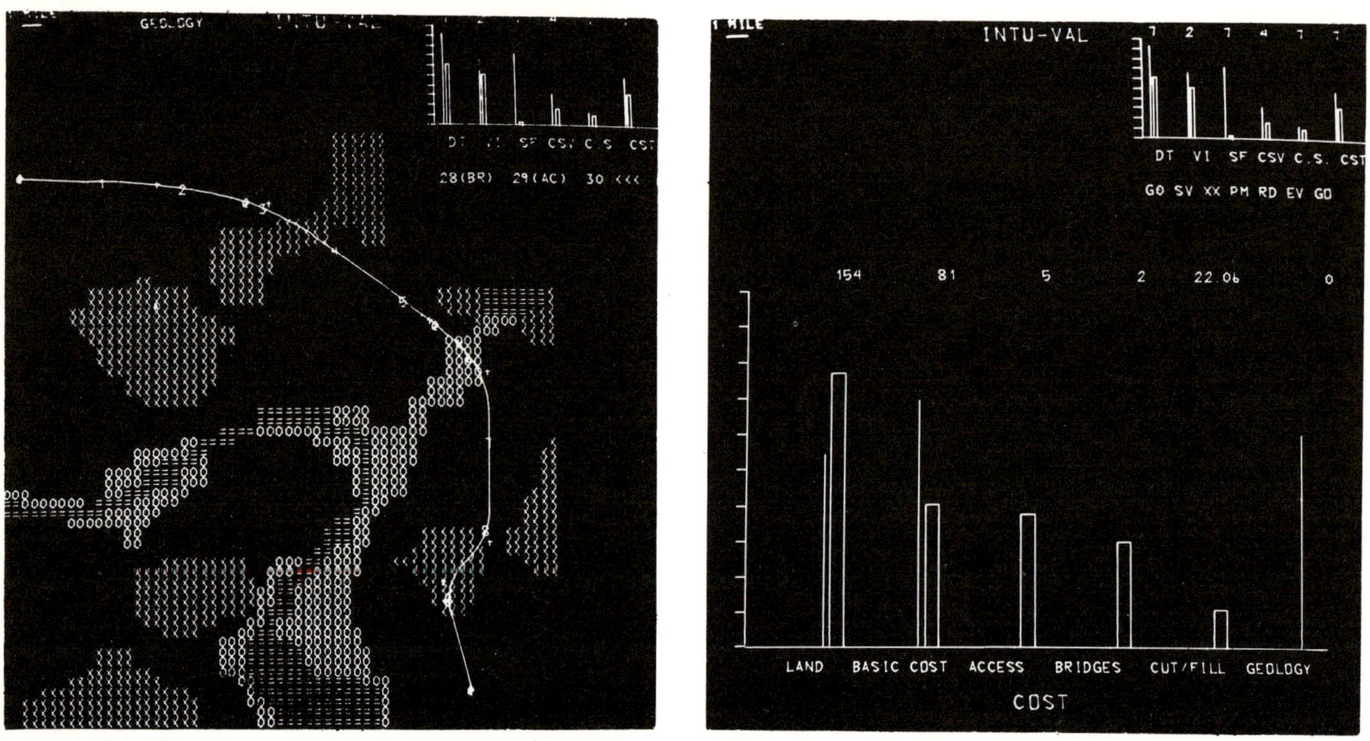

Figure 9 Use of a computer system to answer a complex question – INTU-VAL, a system for exploring land use and transportation planning problems (under development by Peter Kamnitzer).

SIMULATION AND SOLUTION TEAMS IN ARCHITECTURAL DESIGN

Thomas W. Maver

Ackoff (1969), in the context of planning, defines the ideal state as one in which every individual can obtain whatever he wants and in which he has a continuously expanding set of desires. The necessary and sufficient conditions for this state can be listed as:

1) Politico-economic (PLENTY) – to provide every individual with instruments that are perfectly efficient for his objectives.
2) Scientific (TRUTH) – to develop instruments and identify means which are perfectly efficient and to provide every individual with a knowledge and understanding of these.
3) Ethico-moral (GOOD) – to remove conflict within individuals and between them to provide peace of mind and peace among men.
4) Aesthetic (BEAUTY) – to enable every individual to enlarge the range of his objectives through conceptualisation of new desirable states.

By definition an ideal state is that which is unobtainable and is approachable without limit. It is necessary, therefore, to define also a system of objectives (which are defined as ends which are attainable though not necessarily within the period planned for) which, if we are lucky, will give rise to a system of goals which are surely and predictably attainable.

For an objectives system one can take that developed by the Building Performance Research Unit and described by Markus (1967). The system comprises four sub-systems – the building system, the environment system, the activity system and the objective system (Figure 1). The overall objective can be stated as the optimisation of the return on the investment of the client's resources, where resources investment is measured in terms of the cost of providing and maintaining a built environment and return is measured in terms of the activity performance indices.

The degree to which this objective is attainable, its compatibility to the system of ideals already stated and the extent to which it promotes definition of goals is, in essence, the subject of this paper.

Architecture is distinguished from other areas of design endeavour by three main characteristics:

a) the magnitude of the 'solution space',
b) the multi-variate nature of the problem, and
c) the temporal variation in requirements.

Inability to come to terms with these characteristics has necessitated:

a) a retreat into stylism,
b) a hierarchy of priority weightings personal to each architect, and
c) an inflexible monumentality.

It is the intention of the paper to outline a mechanism designed to obviate these shortcomings by:

a) use of the computer,
b) participation by users, and
c) continued use of the mechanism.

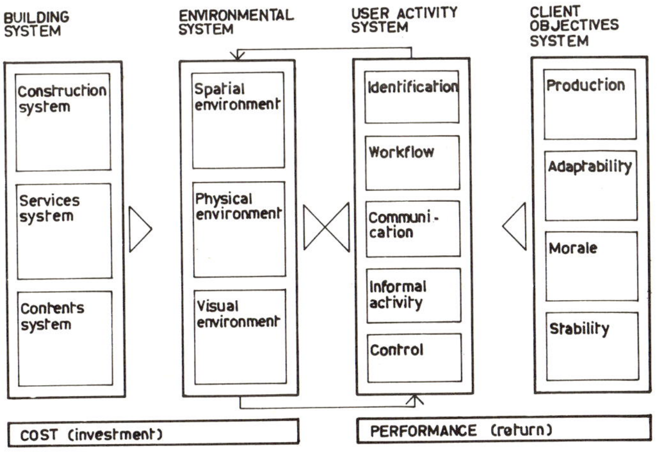

Figure 1 Cost/performance model.

The mechanism proposed, for each stage of the design morphology, is one in which a suite of appraisal programs, covering cost and performance variables, would be applied to a simulated solution hypothesised by the architect; the outcome of the appraisals would be considered by a 'solution team' (as opposed to a design team) together with the non-quantifiable variables; if the balance between cost and performance or between different aspects of performance is not considered optimal, the simulated solution would be modified and the process repeated (Figure 2). The principle embodied is that of making the consequences of design decisions explicit to the 'solution team' which can be composed of client, users, financiers, representatives of society at large, etc. To assist the solution team to negotiate an optimum solution through trade-offs and compromises, the norms, optima or statutory constraints for each variable can be provided as a basis of comparison for each variable performance index.

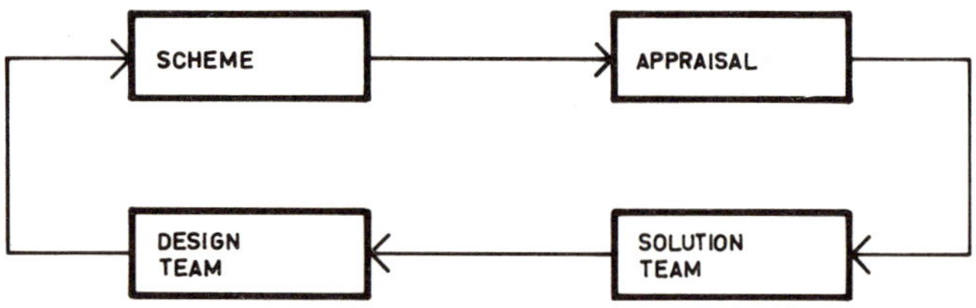

Figure 2 Suggested mechanism.

THE SIMULATION

Figures 3 and 4 show, respectively, the input and output of a computer package, known as PACE 1 (Maver, 1971). The package, applicable to the strategic 'outline proposals' stage in design, elicits from the designer a description of a proposed scheme under five main sections (Figure 3):

General Information:	building type, size, location, etc.
Geometrical Information:	shape, size, location and orientation of individual components.
Site Information:	a numerical analysis of the site conditions.
Construction Information:	areas of fenestration and insulation standards.
Activity Information:	matrix of the functional inter-dependencies between components.

The output (Figure 4) is recorded under four sections and three heads. Under the first head is recorded the absolute value, under the second a unit value (obtained by dividing the absolute value by the number of occupants) and under the third head is the mean unit value of all previous schemes of this type which have been appraised by computer.

The first section of output deals with capital and running costs; the second section deals with the spatial relationships of the scheme to the site; the third section sizes the engineering services plant necessary to provide an adequate environment and, if desired, provides month by month details of the thermal environment; the final section comprises a matrix of standardized values representing the degree to which the arrangement of spatial components satisfies the values expressed in the matrix of functional interdependencies in the input – a high positive value representing too great a distance between components, a high negative value representing too great a propinquity.

```
******
INPUT
EXAMPLE
******
GENERAL INFORMATION
--------------------------

WHAT IS YOUR BUILDING TYPE?
 1=SCHOOL
2=HOSPITAL
3=OFFICE
4=HOUSE
5=FACTORY
6=FACTORY(SHIFTS)
 ?1

WHAT IS TOTAL OCCUPANCY OF SCHEME
 ?1000

WHAT IS LOCATION OF SITE
 1=SCOTLAND
2=MIDLANDS
3=SOUTH
 ?1

WHAT IS THE ALTITUDE TO THE NEAREST 50 FEET
 ?50

***
SITE INFORMATION
----------------

WHAT ARE YOUR X AND Y SITE LIMITS
 ?400,400

HOW MANY ROWS AND COLUMNS HAVE YOU IN YOUR SITE MATRIX
 ?2,2

TYPE IN SITE VALUES FOR EACH ROW
ROW          1
 ?2,5

ROW          2
 ?8,1

*****

CONSTRUCTIONAL INFORMATION
--------------------------

25 PERCENT GLAZING ON 4 WALLS, NO GLAZING ON ROOF AND
MEDIUM STANDARD INSULATION HAS BEEN ASSUMED
DO YOU WISH TO INPUT PARTICULAR GLAZING AND INSULATION
VALUES FOR ANY OF THE ELEMENTS       0/1
 ?0

*****
ACTIVITY DATA
-------------

TYPE ASSOCIATION OF COMPONENT WITH EACH
OF HIGHER NO.
COMPONENT          1
 ?3,5,2,9,8

COMPONENT          2
 ?1,7,6,2

COMPONENT          3
 ?5,2,7

COMPONENT          4
 ?3,3

COMPONENT          5
 ?1
```

Figure 3 Input format.

At the conclusion of the output, three questions are posed. A positive response to the first provides the opportunity of modifying the original scheme; a positive response to the second causes the cost and performance values of the current scheme to be sent up to the data file thus up-dating the mean values for subsequent runs; a positive response to the third

```
*****
OUTPUT
EXAMPLE
*****
                                        VALUE       UNIT VALUE      MEAN
--------------------------------------------------------------------------
COSTS
-----
CAPITAL COST                      £    432000.      432.0000     520.7000
MAINTENANCE COST/ANNUM            £      2851.        2.8512       3.3780
LIGHTING COST/ANNUM               £       387.        0.3872       0.5262
HEATING COST/ANNUM
----------------------
ELECTRICITY                       £      4139.        4.1386       5.4720
OFF-PEAK ELECTRICITY              £      3016.        3.0163       3.9930
GAS                               £      1473.        1.4731       1.9458
OIL                               £      1178.        1.1785       1.5586
ANTHRACITE                        £      1080.        1.0802       1.4296
DISTRICT HEATING                  £      1438.        1.4380       1.9025
WHICH ENERGY DO YOU WANT              1-6
 ?3

HOT WATER COST/ANNUM              £       384.        0.3841       0.3840
TOTAL RUNNING COST/ANNUM          £      5095.        5.0955       6.3045
---------------------------------------------------------------
SPATIAL PERFORMANCE
-------------------
SITE UTILIZATION                                      0.3000       0.3300

SITE VALUE                                            4.0000       2.7250

PLOT RATIO                                            1.8000       1.2500

PLAN COMPACTNESS                                      0.5110       0.7181

MASS COMPACTNESS                                      0.4328       0.4278

---------------------------------------------------------------
ENVIRONMENTAL PERFORMANCE
-------------------------
X-SECT AREA RAIN WATER PIPE             544.          0.5440       0.8905
PERMANENT ART. LIGHT. REQD.           23600.         23.6000      35.8750
MECHANICAL VENTILATION REQD.         472000.          472.          730.
HEAT LOSS/UNIT AREA                                 29621.        27402.
TOTAL WATER STORAGE                   10000.         10.0000      10.0000
SIZE OF HOT WATER CLORIFIER            1000.          1.0000       1.0000
SIZE OF BOILER                      4451615.      4451.6155    5754.8635
FOR HOW MANY COMPONENTS
DO YOU WISH HEAT GAIN/LOSS DIAGNOSTICS
 ?2

WHICH COMPONENTS
 ?1,6
                                82371440.1
                                82194518.1
                                74065774.9
                                58540744.5
                                36146070.9
                               -17325555.5
                               -17146815.5
                               -16724967.5
                                36341946.9
                                48191099.7
                                71851713.3
                                78817386.5


                                39510219.7
                                39173343.7
                                35024742.1
                                27803938.9
                                17661246.1
                                -5629789.1
                                -5441329.1
                                -4765897.1
                                18627762.1
                                24271196.5
                                34729813.3
                                37988794.9

---------------------------------------------------------------
```

Figure 4 Output format.

question produces a paper tape from which a selection of perspective views can be produced (see Figure 5).

In short, the package provides:

a) explicit, numerical cost and performance measures on a range of variables,
b) the opportunity to modify the original scheme or try a wide variety of schemes extremely quickly,
c) a constantly up-dated data bank of cost and performances indices from previous schemes,
d) a visual representation of any scheme.

It is thus possible to place before those who will be affected by the design decision an evaluated range of alternatives.

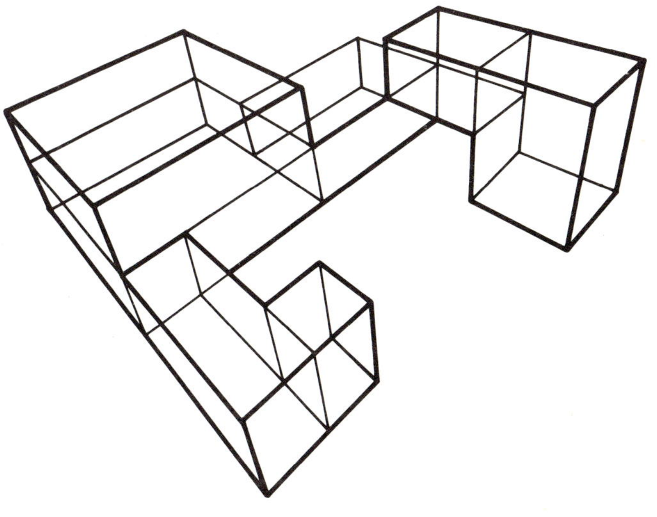

Figure 5 Computer-drawn perspective of scheme can be used as a basis for fully-worked perspective drawing.

THE SOLUTION TEAM

This section of the paper is concerned with a formal mechanism whereby a disparate group of people, all of whom have a stake in the outcome of the design decision but hold different priorities and objectives, can effectively coalesce towards an agreed design solution.

This group of people, say six, are known as the 'solution team'. In order to express their views about, say, five alternative schemes, each is given five votes to cast as he thinks fit. He may cast all his votes for a single scheme, cast a single vote for each and every scheme or cast his votes in any distribution between these two extremes. When the votes are cast (in secret), they are totalled and members of the solution team informed of the outcome. The team members are then invited to re-vote and again the totals are fed back. This process continues until two consecutive 'passes' of the game produce no changes in the voting. This indicates the point at which the members of the team are (apparently) as content with the result as it is possible for them to be, within the rules of the game.

Figure 6 represents the operation of the mechanism; in this case, five passes were performed to ensure coalescence. It will be noted that neither of the schemes with the highest score at the conclusion of the first pass was the eventual 'winner'.

The data yielded by such a mechanism on the way members of the solution team conceive of trade-offs, offer a life-time's work for the committed psychologist. To date, few examples have been tried but in all cases there is an apparent coalescence which gives some hope for the validity of the mechanism.

An interesting development would be the introduction of a 'non-solution' alternative (cf. Ackoff's "John Doe" Presidential Candidate). If the non-solution is the winner, or receives a pre-determined and significant number of votes, this represents a failure of confidence by the solution team in the real alternatives and the *design* team is obliged to generate further schemes or abdicate its role.

CONCLUSIONS

Section 2 of this paper describes a mechanism which promotes, by rapid interaction between computer and design team, effective search within the vast range of possible building design solutions. Section 3 outlines a mechanism whereby the hierarchy of priority weightings of each person concerned with the outcome of the design activity can be reconciled. The two mechanisms combined, operated continuously throughout the life of the building by the users, go some way towards coping with the third architectural characteristic – the temporal variation in requirements. The concept of "design-in-use", made possible by interactive computing and user involvement is one of the most exciting notions to emerge from the current re-think of the design activity.

The mechanisms described in this paper are admittedly crude. Work is proceeding at Strathclyde, however, to increase the range of appraisal measures to include more personal performance variables such as privacy and to monitor solution-team responses. Notwithstanding the crudity of the current mechanisms, they do, I would maintain, promote the achievement of the four necessary and sufficient conditions of the ideal state listed at the beginning of this paper. Their use in practice, however, depends on two major requirements: the resignation by today's building designers of their jealously held role of "value-judge";

SOLUTION TEAM MEMBERS	SCHEMES A	B	C	D	E
1	3	2	0	0	0
2	2	2	0	0	1
3	0	0	5	0	0
4	0	1	0	4	0
5	1	1	1	1	1
6	2	1	2	0	0
Total	8	7	8	5	2
1	3	2	0	0	0
2	2	3	0	0	0
3	0	0	5	0	0
4	0	3	0	2	0
5	1	2	2	0	0
6	3	0	2	0	0
Total	9	10	9	2	0
1	4	1	0	0	0
2	1	4	0	0	0
3	0	0	5	0	0
4	0	5	0	0	0
5	0	3	2	0	0
6	3	2	0	0	0
Total	8	15	7	0	0
1	5	0	0	0	0
2	0	5	0	0	0
3	0	1	4	0	0
4	0	5	0	0	0
5	0	5	0	0	0
6	0	0	5	0	0
Total	5	16	9	0	0
1	5	0	0	0	0
2	0	5	0	0	0
3	0	1	4	0	0
4	0	5	0	0	0
5	0	5	0	0	0
6	0	0	5	0	0
Total	5	16	9	0	0

Figure 6 Solution team voting mechanism.

and the acceptance of the computer, not as a decision-maker, but as a necessary aid for turning decision-making over to those whose right it surely is.

I am one who believes that participation in design decision-making by those affected by design decisions is one of the most fundamental civil rights. If it is to be secured as such, for subsequent generations, we must establish the means now. Our concern, then, must be to design solution-generating systems rather than solutions; to paraphrase McLuhan – THE PLAN IS THE PROCESS.

A DOUGHNUT MODEL OF THE ENVIRONMENT AND ITS DESIGN

Thomas A. Markus

Systems thought has been deeply influential in planning and architectural theory in recent years. Systemic models of cities, buildings, transportation networks, engineering services, building processes and design processes abound. It is commonplace that for any such model to be reasonably useful – i.e., to represent and interpret the real world in a way which has some predictive power – the sub-systems and parts of the model must be properly related to each other. To be related there must be some consistency of definitions, units, concepts and dimensions.

It is quite clear that such consistency is almost totally lacking in the various models which, between them, are supposed to represent what has come to be called the built environment. Some models describe concrete objects in physical terms. Others describe energy systems as thermodynamic systems. Others describe the people who use the systems in behavioural psychological terms and the process by which the systems are designed and built in highly abstract decision-making production-flow terms. This paper attempts to describe buildings, their environments, their users, their designers and the total resources which go into their design, construction and continuing use as parts of a single system. It shows that such a model has important social and political implications.

THE SYSTEM

The system here described is basically derived from the model developed by the Building Performance Research Unit (1967) and Markus (1967). This system is considered to consist of people and things, interacting in a complex way. The 'things' specifically of interest are those pieces of hardware which generate the environment. People are assumed to be goal oriented – seeking to achieve objectives of an idealised kind by achieving more immediate goals. One of the goals which people clearly wish to achieve is a good state of the environment – that is, one which is helpful in achieving all their other personal, social organisational goals and objectives. The system includes certain resources of energy, skill, material and time available for the achievement of its goals.

The system can exist at any scale; for instance a city region, a town, a building or a single space within a building. At any level a system of a smaller scale can be seen as a sub-system or component. The model described below could be developed for any scale, but the empirical research upon which it is based was carried out at the scale of a single complete building and its users. The users were assumed to be members of an organisation and therefore the description of some of the elements in the model was influenced by organisation theory.

The system has five main parts:
1) The objectives system;
2) The activities system;
3) The environmental system;
4) The building system;
5) The resources system.

It is diagrammatically shown in Figure 1 that these five, with their sub-systems and components, make a complex system which is of course open – to influence of politics and economics; culture; climate; the city plan and the site; the social and business context. It is within these that the building universe exists – and it is only to make discussion easier that the model described below is explained in terms which isolate this system from the larger universe.

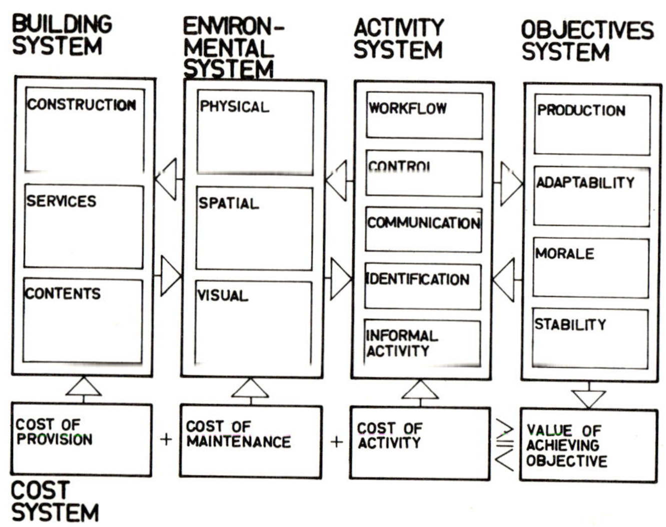

Figure 1

The Objectives System

It is assumed that individuals are goal oriented in order to achieve objectives and as a consequence it is necessary to consider the goals as part of the objectives system. The objectives system consists of those long term aims for which the organisation exists. These objectives provide the context for all the activities and hence for the buildings and environment. Often an organisation's objective is in conflict with broader, social objectives (e.g. industry and pollution) or with narrower personal ones (e.g. production and friendship formation). Such micro and macro conflicts are inherent in all organisations and the designer has to understand them and adopt priorities. Often his own objectives will cause further conflict.

Whilst we are still a long way from being able to specify exactly the objectives of many organisations and isolating those for which the environmental system is particularly relevant, there are four general objectives common to most organisations which are likely to be of relevance to the design process.

Production The great majority of organisations change some resource from one level to another; they create a product. In industry this is obvious; however, the implications of the production objectives are not so obvious in so called non-commercial organisations of which schools, hospitals and houses are examples and as a consequence the more obvious building implications for that productivity are sometimes missed.

Adaptability A slightly less obvious organisational objective is that of being able to adapt. Survival is based upon a two way process of adaptation and for an organisation the ability to change itself in response to changes in the environment is a crucial one. As with other organisms it is likely that also with organisations some of the most crucial limitations on adaptation are set by physical structure.

Morale It seems reasonable to suggest that many organisations have as a distinct objective the wish to keep their members happy. With such non-commercial organisations as clubs this is obviously the case but many industries also claim this is an aim in its own right without the ulterior motive of increasing production.

Stability The turmoil and constant variation which the above three objectives either create or deal with, inevitably give rise to difficulties within the organisation in terms of its stability or the degree to which it exists as a single entity over time and space. As a consequence a further organisational objective will be to maintain the organisation in a stable state so that although production is being maintained or increased, adaptation is taking place and morale is also maintained, the organisation continues to exist in a recognisable form.

These four sub-systems of the objectives system all interact in a variety of ways, this being one of the reasons why they can be regarded together as a system, and between them and their interactions most of the overall objectives of most organisations can be accounted for. In many organisations these require a building or a specific type of environment if the organisation is to move towards achieving them, so it is valid to think of this system giving rise to the need for the further system of the building. On the other hand, the reason for the building is that it generates an environment required for the activities needed by the organisation to achieve its objectives. In other words, the objectives give rise to the activities which it is necessary to implement in order to achieve those objectives. Thus the activity system may be regarded as a set of goals, the achievement of which leads to the reaching of objectives. Objectives are therefore the beginning and the end of the whole system; its vital centre.

The Activity System

If an organisation is described it is usually in terms of what it does; the activities it undertakes; the behaviour in which its members partake. The potential range of these activities is as wide as the range of human potential for action and the way in which these activities are classified depends on the particular aim of the classifier. The purpose at present is to describe as simply and as briefly as possible the range of activities for which the building and its environment are relevant. Five categories are used. These categories do not contribute in direct way to each of the objectives but rather the objectives are achieved as a product of their interactions.

Workflow One of the activities central to most organisations is that associated with modification of resources to give rise to a commodity of greater value. This workflow activity is usually considered the essence of an organisation. A factory is described as 'making cars', for example, a school as a place where children 'are educated', and so on. Many organisation and method studies concern themselves solely with the improvement of workflow activities as these are often the most obvious determinants of organisational success.

One productive process which is required is the construction of environmental hardware – building. Like other specialised processes, for instance catering or laundering, or making machines, this task is frequently contracted to an independent producer.

Control In order to keep the workflow activities continuing smoothly, helped by all the ancillary organisational activities, fed with the right materials and relieved of products at the correct time it is necessary for some parts of the organisation to be responsible for coordination and control of the whole activity system.

Of course, control does not relate only to workflow activities; as organisations become more complex so more energy is spent upon controlling the other aspects of activity. In fact one of the critical growth points of an organisation may be thought of as that point at which it needs to instate processes that are specifically geared to controlling existing control processes, in other words, when a division between senior and junior management takes place. This is no place to discuss the subtleties of this process but it should now be apparent that the relationship between control processes and the other processes which make up an organisation is often critical to the survival of that organisation.

One important goal which any organisation has to achieve is the control of its own environment. This involves the continuous adaptation of space, site and services for full use; planning for replacement, obsolescence and repair; re-organisation of activities in accordance with physical constraints. This activity, in complex situations, is now a specialised form of control commonly called design.

Communication It is not possible to think of the organisation as a static thing. Even to continue to exist at one level it must take in resources and modify them and dispose of those which it cannot assimilate. Many organisations are constantly developing and changing. It is of the essence of organisations that some of their energies are spent transmitting resources, products or phenomena from one place to another. This process of transmission from place to place may be taken as an inevitable counterpart to the process of change from one state to another. The transmission aspect of the activities are referred to as communication.

Communication is taken to include, in most organisations, the movement of people, things, energy, and information. From the two aspects mentioned above it will be clear that the movement of resources through the workflow process involves communication and that the transmission of instructions from control to workflow centres also involves communication; but so will all other aspects of activity.

Identification When communication takes place it consists of transmission of something (or someone) from place A to place B. That much is obvious. What is not so obvious is that it must be possible to identify B as B if communication is to reach the destination for which it was intended. The very essence of the distinction between place A and B is that they have separate, distinguishable identities. If there is any choice at all in the route which the communication can take when it leaves A then the identification of B becomes critical in determining whether the communication gets there. If there is no choice in the route from A to B then in what sense is it meaningful to think of them as separate entities?

We tend to think of identity (especially of people) as something which is part of an object (or person) and which makes it unique. However, in many cases it would seem that a more fruitful way of thinking of identity is as those aspects of the object in question which indicate how it interacts with the other objects in the system of which it is a part. The identity of an object thus relates to its role in a particular system. If the system changes so does its identity.

Out of all this grows the need for an organisation to devote some of its energies to specifying and maintaining the identities of its component parts although in many cases these energies will not contribute directly to the workflow.

Finally it should be pointed out that just as the parts of an organisation must be identified if they are to function adequately, so organisations must have appropriate identities if they are to function in the larger system of which society consists.

Informal activity Not everything which goes on within an organisation is part of the four processes described above and not everything which takes place is directly a part of formally organised activities, or under the control of the controllers. Therefore, a complete description of the activities which an organisation needs to achieve its goals must include a category for this informal activity. The simplest way of thinking of this category is as a miscellaneous one, the size of which is directly related to the sophistication of the organisation or our knowledge of it. This category might also be thought of as containing activities which the organisation needed to deal with its own inadequacies – slack introduced into the system in case the strain grows. This latter possibility seems the more plausible on the basis of a model of people as goal oriented. If they are goal oriented it is probable that they have personal goals in conflict with those of the formal organisation. Achievement of these personal goals may well be essential for morale, however, and slack in the system may serve to make them possible by informal activity.

One description of informal activities, then, might be those activities brought about by a mismatch between the goals of an individual and the goals of the organisation of which he is a part. In allowing individuals to find ways of satisfying their own goals within the organisation they are in fact encouraged to continue as members of the organisation. A further point worth considering is that an organisation develops within the context of a particular economic, social and political climate and as a consequence might not have built into it the possibility for coping with changes in that climate. However, individuals are not tied in quite the same way and hence their informal activities can contribute considerably to the organisation's ability to adapt to change.

This means that provision for informal activities will enable the goals of individuals and of society to be achieved. It raises the whole question of the designer's responsibility for, and open-ness to, values and goals other than those with which he is formally presented. It is also relevant to ask to what extent deliberate planning and provision for informal activities by authorities is a self-defeating process.

The Environment/Activity Interface

If it is to function properly, any activity system must have an appropriate environment. In most cases this appropriate environment is provided in buildings, which modify the external environment in various ways and provide a controlled, internal environmental system within which the activity system can flourish. The relationship between these two systems is particularly intricate and the elucidation of this relationship is central to the development of an understanding of building and environmental design.

People are active and consequently they modify their environment in order to change the way it affects them. This mutual interaction goes on constantly, and it means that it is incomplete to consider an environment without an activity taking place within it or vice versa.

The Environmental System

The environmental system is required to facilitate the activity system. Different activities require different environmental levels and different ranges of adjustment; these are found within different buildings and within different parts of the same building. It is important to distinguish between buildings and the environmental systems they provide because two similar buildings can produce quite different environments. Organisations build not because they *need* to build but because they require a particular set of environmental conditions for their activities. These conditions must be variable within the limits required by the people carrying out the activities, and, of course, they must control unwanted variations in the external environment. People modify their environment according to their activities, sometimes deliberately by adjusting or making alterations, as when they switch on lights, sometimes accidentally, or at least unwillingly, as when electric light increases the heat level.

Categoric distinction between the sub-systems of the environmental system is difficult because of their highly interactive nature, but two recognisable sub-systems clearly emerge.

The physical environment Those aspects of the environmental system directly perceived by the senses: heat, light, sound, texture and smell.

The spatial environment Those aspects of the environment related to the dimensional and geometrical properties of single spaces and to the spatial relationships between them.

These two are very broad categories; but it is clear that interactions with activity is high. For example, noise and light affect work output and communication; people continuously adjust, extend, rearrange their spatial environments.

Visual environment One characteristic of environment is of exceptional importance and straddles both sub-systems. It is the visual quality of spaces seen in light. This is central to the art and practice of architecture and is the vehicle whereby meaningful experience is obtained from form. In the design of the building system many choices are made for the sake of this characteristic. Its experience is aesthetic experience, one of the human goals motivating the activity system even though the production of satisfactory aesthetic experience may not be explicit as an organisational objective. It is all-pervasive, throughout the system.

The Building System

In order to achieve its objectives an organisation requires a certain environment and to achieve this environment it produces a building. The stuff of which the building is made, the bricks-and-mortar, components, service installations and so on, comprise the building system. It includes all those items normally described in drawings, specifications and bills of quantities and all tangible contents other than human occupants.

The building system gives rise to the environmental system by modifying the external conditions, and this modification is done in two distinct ways. Firstly, by exclusion, or filtering or selectively admitting through the fabric and secondly by consuming energy to generate an environmental condition.

Three sub-systems of the building system can be readily described.

The constructional sub-system Within this are categorised not only the structure, be it frame, shell or whatever, but all the inert, not directly energy consuming, constructional parts of the building fabric.

The services sub-system The service installations concerned with the supply and disposal of water, gas, electricity and fluids and solids for use in the activity system or in the modification of environmental conditions.

The contents sub-system Plant and equipment, furnishings, fittings and finishes. Precise definitions of the distinction between the previous two sub-systems and this last is difficult, and perhaps the best working definition is that the contents system comprises all the hardware of the building system not included in the two previous sub-systems.

The Resources System

Each of the four systems described above has an initial and/or continuing cost or value. The building system costs something to provide. The environmental system has costs of energy maintenance, cleaning etc., associated with the maintenance of any given environmental state. The activities consume resources – wages and salaries, materials (used and wasted); advertising; recruiting; image-making, etc. The objectives have values. These values should exceed the combined cost of the first three systems – otherwise the system is running 'at a loss'. The difficulty of quantifying values in cash, or other units commensurate with costs, should not blind one to the need for, and the possibilities of, adopting cost-benefit analyses for many design problems.

DESIGN AS A BASIC HUMAN ACTIVITY

The activity of design is a purposeful, goal-oriented search. The search is for a physical solution to a perceived and, more or less, understood problem. People have perceived such problems from the earliest times. Sometimes survival depends upon a successful outcome to this search – say a search for safe shelter. Often the problems are only dimly perceived; sometimes they consist of a whole group of related problems whose complexity is hidden by the apparent simplicity or unitary nature of the solution. Always the search has to be successful within certain constraints; time (before the cold weather); energy (not too far; or not too heavy for two men to lift); skill (capable of execution by a limited technology); money (purchasable with the available funds).

Human development has often been described in terms of design and technical ability and periods are conveniently labelled according to the technology adopted in them. Thus design can be regarded as a fundamental human activity requiring both consciousness and thought to understand the environment; will and purpose to control, change or improve it; abstract thought to imagine changed states of the environment in anticipation; and skill to bring intentions and plans to concrete realisation.

So, more formally, one might define design activity as; action aimed at finding solutions to perceived problems within a resource envelope. However, since such a definition covers any action aimed at solving problems, even where the solutions are decisions related to personal or organisational behaviour, say a decision to fight a battle; or to understanding a set of mathematical equations, a further refinement of the definition is needed. For the purpose of environmental design in general, the solution in part, at least, must consist of physical systems – some hardware. This hardware affects environment and thus the lives of individuals and organisations existing in this environment. It is part of an interactive animate/inanimate system described earlier. One might also usefully add that further, productive action results from decisions about these hardware systems.

Above, a model was proposed which relates in one integrated system, people, things and people's resources. In the past, designers' descriptions of buildings have generally been in hardware or environmental terms, and the effect of these on people has had to be described by inventing a special species of people called 'users'. Their goals have been called 'user needs' and much fruitless survey work has resulted from the lack of an empirical model which related things and people.

Behavioural scientists, on the other hand, have traditionally regarded the variables of physical environment as intervening nuisances which have either been held constant or, more often, ignored. Professional designers forgot that it was only recently that the magnitude of design problems has caused them to be employed to carry out what has always been a communal activity; and behavioural scientists have overlooked a tremendously rich field of observable action which contained information on values, imagery, social networks and concepts.

CONTROL OF WORKFLOW

The interactive systems model described earlier contained, as one characteristic organisational activity, control; the kind of generative, organising acts which relate all other activity towards objectives and goals.

Organisational theory has generally seen control as being concerned with other activities – 'design of activity patterns' might be a useful description. But one important aspect of control is the control of productive activities (workflow) for the creation or continuous re-creation of environments best matched for human purposes. The outcome of such productive activities are new or changed products – pieces of hardware (buildings, say). Another activity shown in the model is workflow; that is, productive activity. The production of buildings and other environmental hardware is just as much part of the workflow process as design is of the control process. It is similar in another respect too; that is that it is often delegated to a specialist production unit – e.g., a contractor. So design and production of hardware is part and parcel of the systems model. And the achievement of the best state of the *whole* system represented by the model can be seen to be an ever-present objective.

An important part of the whole system is the resources system and to say that the best state of the whole system is an objective is another way of saying that the best allocation of resources in that whole system is an objective.

Thus we have *inside* the system two activity sub-systems – building design and production – whose objective is a state of the *whole* system. Design is a generative sub-system which enables the larger system to exist, change and remain whole. The paradox of a sub-system being the generator of the system of which it forms a part – being inside as well as outside it – is conceptually similar to certain topological problems in mathematics in which surfaces of holes in objects continue to envelope the entire object (the hole in the doughnut).

For confirmation that design and production of environments are present in most social or organisational systems, one has only to look at a few examples. Individuals, families and tribes, design and build their furniture, houses and settlements. When the problem concerns complex organisations and technology a simple discussion between the parties (say parents in a family) and production by 'do-it-yourself' may not suffice. Although in some primitive communities design of dwellings and settlements is still a public act, done by the community, specialised designers and producers are usually employed today.

POWER AND DESIGN

What effect has this professional role of designers, and specialist role of building producers, outside the community which becomes the end-user of the created environments? The simplest answer is that environmental control – design – has become as much part of authoritarian and bureaucratic processes as all the other activities controlled by the 'controllers'. This control group hires and fires in-house or independent professional designers who will give them the environments which best suit the organisation's purposes. Certainly some built-in design freedom is tolerated in the system – as part of 'informal activity'. A student may put pin-ups on his study/bedroom wall; a tenant select his own curtain material or front-door colour.

At a subtler level it has become clear that if design is, indeed, such a potent part of the whole system, conflicts about who designs, and what is designed are likely to be amongst the first issues in many revolutionary movements. The destruction of property in riots, and vandalism, can be seen as negative production. The taking over of buildings, their re-use for emergency purposes and ultimately their re-design has always been necessary when power shifted. The polarity between authority and individuals or groups can be seen in environmental conflicts of many kinds:

1) Organisations value a concrete image; unified, purposeful, concentrated. Individuals often value a more ambiguous, diffuse and varied environment (e.g. workers in shirt sleeves and braces were seen as violating the modern and neat image of the C.I.S. offices in Manchester).

2) Organisations generally oppose ad hoc environmental adaptations by their members.

3) Subversive organisations generally work without formally dedicated environments; when the Christian Church or the Communist Party adopted such formal power symbols their revolutionary message was already waning.

4) The control of expenditure on building is always separated from other resources – housing cost yardsticks; pupil-place costs; cost/bed in hospitals. All attempts to include these resources with, say, cost of wages or materials are strenuously resisted.

THE POSSIBLE ROLES OF DESIGNERS

It would appear, then, that design, as a generative and potent sub-system of all human activities, can initiate wide-ranging social and political change.

Western style democracies have realised this and there has been an immense flow of resources into research and development into planning 'participation'. User-need studies; public enquiry systems; advocacy planning; participatory design – these are but a few of the activities which have come about in response to genuine democratic pressures. The literature suggests, however, that they are examples of Schon's 'dynamic conservatism'; ways of appearing to change but actually staying still. Appeals to the national interest, to the expertise of the planners and designers and to long-term benefits have been used to justify environmental inequities of the grossest kind. Many planners and architects have felt themselves to be in the vanguard of liberal reform but have not faced the fact that they mostly have to be employed by authorities who will ultimately resist any reallocation of power.

They have, now, three main alternatives:

1) To continue and increase emphasis on professionalism of a kind which is based on 'expert' and 'inspirational' roles. Such a role depends on stable social structures, material and social recognition of design skills, together with the full system of legal protections. Such a designer, whilst protesting his independence of political pressures, can only function by public or private patronage from the centres of power. Broadly, he is a conservative.

2) To adopt a sympathetic stance to so-called 'participatory' design processes. Such a designer is likely to accept the growth of his professional body to include new disciplines and educational backgrounds. He will work less as a value-judge and be concerned to make the design process more transparent. He will readily adopt design games, computers, public simulations, etc., as means of developing *ranges* of possibilities for public choice. Whilst this range is still his decision, it is likely that solutions will emerge which effect some compromise between authorities and planners on the one hand and the planned on the other. Such a designer is generally from a middle-class leftish background – he rarely comes from the communities for which he works nor lives in them. Broadly, he is a liberal reformer and is likely to have a self image as a democratic community leader.

3) To reject both above solutions and work for a real transfer of power on design decisions. Such a designer will seek employment by clients who will be end-users; but generally will find his work will have to be voluntary as such groups control no environmental resources (e.g., tenants in twilight housing; factory workers; hospital patients). He will see his role as one who has powers of *discerning* the latent solutions already existing in the patterns of life and values of his clients. These he will try to develop, but his expertise will be more that of a midwife. He will generally reject participation in this process, as he may claim that this is a) too sensitive a matter to be capable of resolution by crude group processes, and b) a technique of political manipulation.

He will be much concerned with the language and concept barrier (of Bernstein) which separates him from the majority of the population. He will try to live in his work-environment. Broadly, he is a revolutionary.

Of course there is a whole spectrum and not three categories. But the key to where a designer lies along this spectrum is his view of the relationship between environmental control and all other control in the system.

MODELS OF DESIGN

Many models of design have been produced in the last fifteen years. None of these, so far, has started off with the social and political status of the designer and hence all have failed to relate the design systems they represent to other social and political actions. Not surprisingly, since real design is a socially significant and, largely, socially controlled activity, these models have failed to convince designers of their realism or significance. Systematic design conferences and discussions have become fragmented by a relatively esoteric, small group, too often 'eunuchs' in terms of design experience. Refinements of these models – in terms of feedback loops, spirals, trees, networks etc., are, of course, important. But they cannot succeed in adequately describing design unless they are rooted in careful social analysis; even less can they hope to be formative in educating a new generation of designers.

CONTINUOUS DESIGN AND PRODUCTION

Initial design and continuous re-design are not categorically different activities. The main difference is that in the former only half the system – objectives and actual or proposed activities – is present: the necessary environmental hardware comes into being in conceptual, abstract, form which can be modelled; whereas in the latter, a whole system exists although alterations to it still have to be modelled before being produced. Today professional designers are employed for the former but not, usually, for the latter; this does not mean that continuous design and production of hardware does not go on. Studies amply demonstrate that they do. If the initial designer has done his work well he will leave behind:

a) A hardware system which is capable of being adapted (i.e., robust), and
b) A design or decision-making technique which is not only sensitive to the feedback inputs and capable of continuously good solutions, but have built into it monitoring devices about the design technique itself and a structure capable of sensing and meeting demands for change in this technique.

Figure 2 shows the system model extended in a third dimension, that of time. A cross section through this solid shows a fixed 'snapshot' at any moment of time, of the whole system, which is dynamic. The

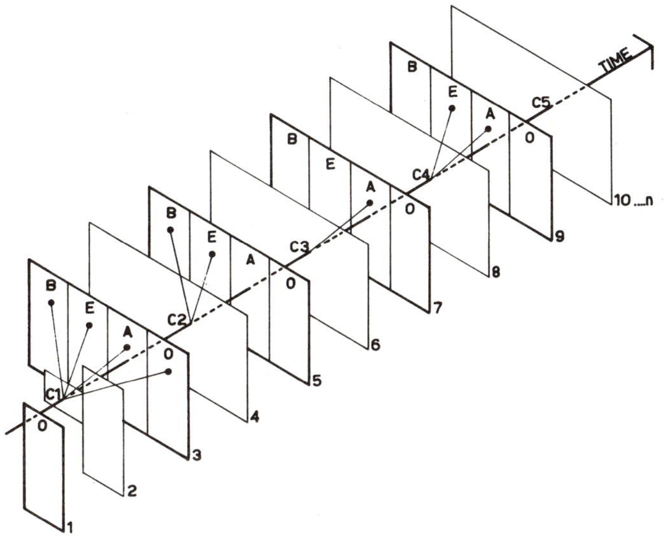

Figure 2 1. Objective system only exists.
2. Objective system and control system exist: design starts, theoretical activity systems, environmental systems and building systems modelled and tested: decision made.
3. All systems exist: dynamic homeostatic balance.
4. Balance upset: parts of system theoretically remodelled and tested: decision made.
5. Balance restored.
6. Balance upset, as 4.
7. Balance restored.
8. Balance upset, . . . etc.

C = control of activity/control of environment = design.

generative sub-system of design ('control') is the connecting link between these layers, which determines the way one changes into the next. The whole of this solid is concerned with the birth, life and death of one scale of system – say a building and its occupants. Above it are solids of greater scale, below it those of smaller scale and all around it similar ones.

This continuing nature of design and production makes it unnecessary to distinguish too carefully between initial and re-design processes. The initial design of any part of this system can be seen as re-design of the system at the level above. Thus a new building in a city is merely an alteration to the city's fabric. A new room, to the fabric of the building; a new window, to the fabric of the room, and so on. City planners, architects, component designers are each part of a continuous dynamic design process. 'Start' and 'finish' are arbitrary operational terms which make designers' labour measurable, but no more.

What now emerges is a generative sub-system which becomes a genuine means of political power. It enables personal and local interests to be reconciled with national and regional ones at a level of *ends* – quality of life, which removes much of the need for experts at the conflict stage; once resolved, experts are needed to bring the decisions into being.

Computers, simulations, new communication media and all the rest may enable us to return to a more primitive *form* of community design control whilst achieving results which primitive technology and society could never approach.

ACKNOWLEDGEMENTS

Many of the ideas in this paper were worked out by and within BPRU. The author has used sections of the BPRU final report but takes personal responsibility for their interpretation.

VALUE THEORY AS A VEHICLE FOR USER PARTICIPATION IN DESIGN

J. N. Siddall

Direct specification of value is the most coherent means of user participation in design. *Value is defined as that quality of a design satisfying the user's desires*, and is an inherent property of an engineering design. The designer must design his product or system so that it generates maximum value for the user. Suppose, for example, he is designing a powered toboggan, illustrated in Figure 1. He will try to give it features such as low cost, good gradeability, long life, good serviceability, high carrying capacity, and low fuel consumption. He does this because he knows *by some means* that the user finds these characteristics desirable. So the designer assigns, usually intuitively, a value to each characteristic or design variable. This value is a measure of the satisfaction that the user will derive from the machine, and each variable makes some contribution to the overall satisfaction provided. The amount of the value will depend on the quantity assigned to the variable. But the designer will also have to trade off variables, and, for example, sacrifice long life for low manufacturing cost. The design process is thus one of selecting the best combination of engineering elements to synthesise a design, and then adjusting the quantities assigned to variables in order to maximise the total value potential.

The value-important characteristics suggested for the example of the powered toboggan are all utility values and the ones traditionally catered to by engineering designers. A good engineering designer would also take into consideration less tangible, but still utilitarian, characteristics such as comfort, convenience and safety. The powered toboggan, commonly called a snowmobile, was, in fact, strictly a utilitarian device when first designed and built, being intended for trappers, hunters, transmission line maintenance men, geologists, and the like. In recent years it has become enormously popular with the general public, and the illustration in Figure 2 suggests that it has more subtle values than strictly utilitarian ones.

Prescription of correct values is the key to successful design. History illustrates (Ogburn and Thomas, 1922) that inventions will occur in response to need if the state of the art is ripe, but technology has failed if the correct prescription of values is not satisfied.

It seems clear that one important mode of user participation in the design process is by carefully and adequately defining his values *directly to the designer.* We shall examine how this might be done, but first it

Figure 1 Powered toboggan or snowmobile.

Figure 2 An expression of non-utilitarian value in the powered toboggan.

will be useful to explore more deeply the concept of value from the viewpoint of biological evolution, philosophy, economics, psychology and market research.

BIOLOGICAL EVOLUTION

It would seem appropriate to begin the examination of the value concept with a study of the role of biological evolution as the origin of at least certain values. This area in the study of values is predicated on the assumption that biological evolution is the mechanism for the creation of the human species. Any aspect of human behaviour can then be examined from the point of view of its origins in the evolutionary development of the species (Berrill, 1955; Lorenz, 1966; Morris, 1969). If a given aspect of behaviour would logically appear to have contributed to survival of the species during its biological development in the pre-cultural era, then it is assumed to be programmed into the genes, and instinctive rather than culturally conditioned. Such an approach provides a very basic mechanistic explanation of modern human values – values that are otherwise sometimes rather difficult to rationalise, and sometimes difficult to even recognise.

Morris (1969) shows, for example, that it has been programmed into our genes as a survival mechanism that a man and woman should adopt pair-bonding, or fall in love. And he shows similarly how the exploratory or innovative urge is an instinct for improvement of survival likelihood of the species, and how it is controlled by a counterbalancing instinct for resistance to change. Following this approach, it can be suggested that man must have an instinctive drive for technology because of its survival value – explaining why man enjoys owning things and tinkering with them and why he has a tendency to develop technology for its own sake. Similar arguments could be used to explain pleasure in games, and pleasure in quiet natural surroundings.

The main conclusion from evolutionary studies is that some values are of biological evolutionary origin, and some are of cultural origin. It may be important to distinguish between them. Instinctive values are difficult, even dangerous, to tamper with.

PHILOSOPHY OF VALUE

While rarely providing concrete aids to decision making, or unequivocal conclusions, philosophy does provide insight into abstract concepts – and few concepts are more important than value. In speculating about value, philosophers attempt to explain its real nature and how it enters into the behaviour of man.

Philosophers begin by attempting to define value – and even this is extremely elusive. Rescher (1969) lists ten different definitions of value by different philosophers, and there are many more. Some, like Spinoza, tell us that values are absolute, given to us by God. Somewhat similar to this viewpoint was that held by Plato, that things have a permanent objective value independent of the user. This value is a fixed property of the thing, which man must learn to appreciate.

Beginning with Aristotle, other philosophers have argued that values are subjective. They are *purely* subjective if they simply reflect a state of mind of the user. They are *partially* subjective if they result from an interaction between a user and an object. Perry (1954) is a modern writer subscribing to the subjective viewpoint. He defines value as follows – "a thing – anything – has value or is valuable, in the original or generic sense, when it is an object of an interest – any interest". He thus associates value with things, although in a subjective way.

Parker (1957), also a modern philosopher, is more purely subjective. He asserts that value resides only in activities or experiences. Objects do not have value: they are only valuable because they have a potential for generating value when they are appreciated or used. Thus one does not desire an object, one desires the experience of enjoying the object.

Existentialists go even further to divorce values from things. Sartre (1965) asserts that values are merely aversions, a means of freeing ourselves or escaping from isolation or loneliness, from pain, danger, annihilation or not-being. Man synthesises a set of values to justify his own existence. This is really a self-deception because man's existence is meaningless.

This very brief discussion will perhaps suggest the importance of how the concept of value is interpreted. A different interpretation could lead to a different sense for the value prescription.

ECONOMIC THEORY AND VALUE

Economists have used something closely corresponding to our concept of value for many years. They call it utility. A typical economist's definition of utility (Seftwich, 1966) is – "Total utility refers to the entire amount of satisfaction obtained from consuming various quantities of a commodity". The concept, in conjunction with hypothetical utility curves as shown in Figure 3, is used by economists to explain in general terms consumer behaviour. Utility units are arbitrary.

The mathematics of decision theory (Von Neumann and Morgernstern, 1944; Horowitz, 1965) is based on the assumption that a person, faced with two or more alternative courses of action (designs), will choose the one with highest utility or value. Or, if uncertainty exists, one will choose the action leading to the highest expected value. It is thus clear that decision making is closely related to the concept of value.

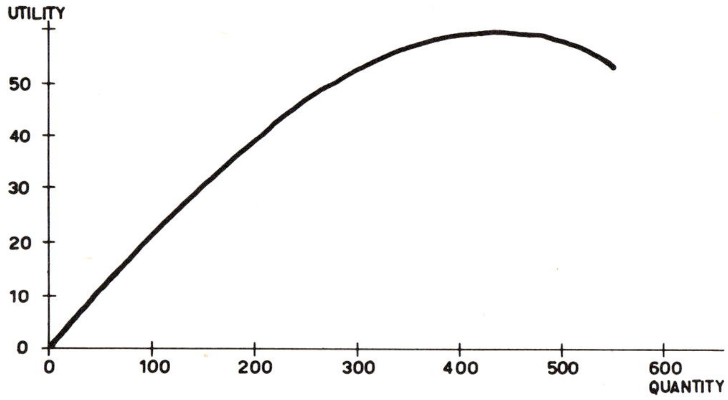

Figure 3 Economist's utility curve for a commodity.

PSYCHOLOGY OF VALUE

Psychologists attempt to explain why people make decisions the way they do. Economists, mathematicians, business scientists, and to some extent philosophers, attempt to say how a rational man ought to make decisions. However the approach of psychologists is far from uniform. "Behavioural scientists attempt to understand what man is in terms of what he does" (Nicosia, 1966). Physiological psychologists attempt to understand what man is in terms of his neurological processes. One approach, for example, suggests that decision making is neurological inhibition of all alternative actions but one (Diamond, Balvin and Diamond, 1963). Another school of psychology tends to dismiss values as not really relevant, and claims that decision making is purely a question of stimulus and response. They assume that we are conditioned by advertising, habit, social custom and the like to respond (make a decision) in a given way to a given stimulus to need. They suggest that no rational consideration of values really occurs. This viewpoint is actually quite close to existentialism.

The behavioural psychologists show us that the decision process tends not to be as straightforward as one might assume. People frequently do not make rational choices based on identified values. Lund (reported by Young, 1968), for example, experimentally examined the determinants of belief or the bases for why people believe various things. The concept of beliefs is closely related to values – beliefs about what is desirable for oneself or society.

In a similar vein, other psychologists present evidence to support the stimulus-response theory of decision making. People tend to often make decisions by habit, following long established patterns of behaviour.

Psychology warns us, then, that designers cannot assume that users will make rational choices paralleling those of the designer.

MARKETING RESEARCH AND VALUE PRESCRIPTION

Marketing research would appear to be closely related to our problem of value prescription. Here, apparently, is a tried and proven technique of determining what products people will buy, how much they will buy, and why they buy them. It would thus appear to be an ideal technique to determine people's values. Indeed marketing research professionals might well contend that no problem exists; that they are in fact regularly and satisfactorily determining people's values *now.*

The initial market research approach was through statistical surveys, using techniques (Phillips, 1968; Siebert and Wills, 1970) such as house-to-house surveys, trial samples, mailed questionnaires, telephone surveys, consumer panels where members record in a diary all purchases, and retail audits where accounting and inventory records are used in sample stores to determine how many of a given product are sold in a given period.

Companies using these statistical surveys discovered that the results were not always reliable; not because of errors in statistical techniques, but because of difficulties with the questions. Ambiguities are extremely difficult to avoid; but even worse is a common tendency of people being questioned to give unreliable answers. This may be due to several reasons – a wish to be polite and please the interviewer by saying one will buy an indifferent product, a wish to appear knowledgeable or gain status by giving fanciful answers, a refusal to admit to socially unacceptable values, or a real unawareness of the true motivation which arises from their unconscious mind. This unawareness may be because the values are actually instinctive but are rationalised in some way, or it may be because of cultural conditioning (Leonhard, 1967).

In order to circumvent all these difficulties, marketing men began to call on the expertise of behavioural psychology, and this led to what is variously called depth research, motivation research and projection techniques. Depth research attempts to determine what motivates people to buy given products, and what their response to advertising is and why. Three techniques are commonly used – the group interview, the thematic apperception test, and invited role playing as a stimulus to free expression (Leonhard, 1967).

An important area of application of motivation research is to the design of advertising (Bliss, 1967) and it is here that it has received its greatest criticism – exemplified by Packard's (1958) well known book "The Hidden Persuaders".

There is no question that mass media advertising based on motivation research is a deliberate attempt to manipulate and exploit unconscious, preconscious, or immature values. Such values do not correspond to those consciously reasoned out as best for oneself, one's group, or mankind generally.

Having gained some insight into marketing research, we now wish to answer first the question – is it now successfully and satisfactorily determining people's values? Marketing research is marketing oriented, and controlled and used by marketing management for the benefit of individual companies, not for society in general. Marketing research asks – what will people really buy, and how can they be persuaded to buy? We are asking, as designers – what do people really desire, and what will best satisfy their desires? One approach is the legitimate concern of business – the other is the legitimate concern of the professionals responsible to society for technology. The two concerns are not necessarily compatible.

The second question to be considered is – can such techniques from the behavioural sciences be adapted to a determination of a value prescription for design, and should they be so used? There is considerable evidence to suggest that they cannot and should not.

1) Depth research does not appear to have made any real attempt to determine the full value profile people associate with a design.
2) Depth techniques probe only the preconscious (Leonhard, 1967; Siebert and Wills, 1970), or middle layer of consciousness, and do not determine values from the unconscious mind.
3) It is admitted by marketing researchers (Leonhard, 1967) that depth techniques do not work well with highly educated people because they realise what the interviewer is up to and tend to reject him.
4) In depth techniques the user is passively being analysed and interpreted. It can be argued that this violates his right, indeed perhaps his duty, to actively, consciously and rationally participate in the prescription of his and society's values.

Although we may reject so-called depth research techniques for determination of a value prescription in design, it would be foolish to reject the possibility of applying the behavioural sciences to our purposes. But it must be done with the active and conscious co-operation of the user, by techniques perhaps not yet formulated – techniques that help the user discover his own values, conscious and unconscious, and integrate them with those of society.

PRESCRIPTION AND COMMUNICATION OF VALUES

We are now ready to consider the problem of how a user may formally communicate his values in a design specification. There are two phases to this – the user must first be shown how to recognise his values, and then how to quantify them and communicate them to the designer.

Some values are hidden in the user's unconscious mind, either because their source is biological evolution and therefore instinctive, or because they are due to cultural pressures since early childhood, either forgotten or repressed into the unconscious mind. Many more values are instinctive than is perhaps generally realised, and such values include the desire for food, drink, sex, health, excitement, entertainment, friendship, power, group acceptance, satisfaction of curiosity, use of intellect, ritual, satisfaction from games, satisfaction from accumulation of material objects, satisfaction from love of technology for its own sake, and ecological satisfaction. Instinctive values are quite straightforward, and not difficult to learn to recognise in oneself. Culturally repressed values are much more difficult, because people have psychological difficulty in admitting to them, even to themselves – as was illustrated in discussing market research. The behavioural sciences should be able to develop techniques to assist the user in recognising such values in himself, and in treating them rationally.

Conscious and rational values are easier to handle. They include things like recording of knowledge, moral values, and utility values. Recognising values, whether conscious or unconscious, is aided by a classification of value categories like that suggested when discussing philosophy of value.

Having identified all his true significant values, the user's next problem is quantification so that they may be communicated meaningfully to the designer. Many utility values can be quantified using utility curves somewhat like those used by economists and business scientists. However, the use of indifference bets to set up a utility curve does not seem feasible, and they are established by direct use of intuition. Lifson (1962) appears to have been the first to have proposed the use of utility or value curves in the design of devices or systems.

Utility can, as we have suggested earlier in this paper, be considered a function of design characteristics or variables such as cost, gradeability, life, capacity, fuel consumption and the like. Each makes its own contribution to value, depending on the quantity assigned to it in the actual design. Returning to our example of the powered toboggan, some of the important characteristics might have utility curves like those shown in Figure 4, if the user is a trapper. Specific curves are valid for any possible design for a given user or users, and for a given purpose. Each curve is quite independent of the value contribution of any other variable, and interactions are ignored. It may be noted that some curves drop to a utility of minus infinity if the design characteristic does not meet some specific requirement, thus killing off all utility of the design.

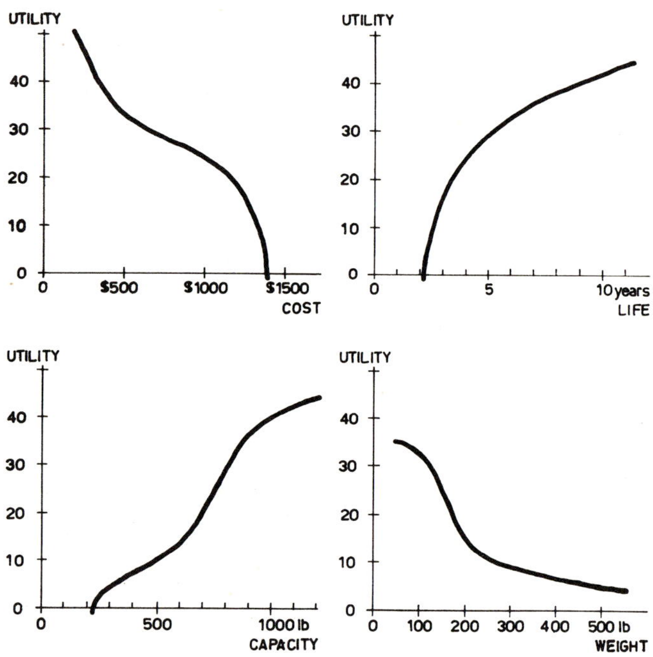

Figure 4 Utility curves for a powered toboggan.

Such curves must be defined *subjectively*, using arbitrary utility scales; and the different curves for a given design must have utility magnitudes reflecting the relative importance of the design characteristics. It may be argued that one cannot put a numerical measure on a subjective quantity like utility. However it would seem possible, with practice, that one could consistently assign a number to represent the degree of a certain kind of desirability. It is similar in concept to subjective probability, which is the degree of an individual's belief that an event will occur, and now widely accepted as a general definition of probability. Since the curves are subjective, it is widely preferable that the user define them first hand, rather than the designer attempting to second-guess them.

Occasionally a non-utility value curve can be defined. For our example of the powered toboggan, ecological value could be considered a function of noise in decibels. However, many non-utility values cannot be conceived of as functions of design characteristics. Nevertheless they can be given a numerical measure *corresponding to a given design*. Thus they cannot be specified with the same generality as when value curves can be used. The user must be confronted with two or more actual or proposed designs, before he can assign value measures for comparison. Important non-utility values for the powered toboggan might be excitement (function of speed), entertainment, friendship (from clubs), game, material, technological, and ecological. It should be noted that some value measures could be negative, notably ecological. There has been considerable complaint recently in Canada that these vehicles disturb the natural environment of wild areas. It is very important that the user prescribe directly his non-utility values, using a check-list of value categories.

COMBINING VALUES

We have seen how a person may subjectively prescribe his values corresponding to a design, but the question remains as to how one combines them to give one overall measure of value. Simply adding them is a solution that leaps to the mind, but it is possible that there are interactions among values when they are combined.

Instead of maximising desirability in decision making, it may be that the intuitive mind, in some people or at some occasions, minimises undesirability. If combined value is represented by U_T, and value components by U_1, U_2, etc., then undesirability may be represented by either $-U_T$ or $1/U_T$ (the inverse concept was suggested by Sutherland, 1970). If we assume component undesirabilities add up, the second concept leads to

$$\frac{1}{U_T} = \frac{1}{U_1} + \frac{1}{U_2} + \ldots$$

or

$$U_T = \frac{1}{1/U_1 + 1/U_2 + \ldots}$$

The question of combining value components has not been wholly resolved.

A second problem in combining values occurs when there is more than one user. How do we determine a consensus of values? Confusion will result here unless we restrict ourselves to the decision of what is the best design, and do not attempt to decide at this stage whether any design at all is desirable.

It is assumed that two simple basic principles underly such a determination – everyone should have the same value share, and no one can veto a given design. The second consideration precludes negative total value for a design – an individual cannot score an undesirable design at less than zero.

We have seen that a full value prescription is, in general, not possible unless specific designs are available to which measures of value components may be assigned. In this event, one would be inclined to suggest that the best choice should be decided by simple voting. However, it has been shown (Rescher, 1969) that, if there are more than two alternatives, voting may not give a correct choice. A proposed procedure would be to take the maximum of each user's value assignment to the different alternatives (say four), as follows

$$\overline{U}_i = \max\,(U_{i1}, U_{i2}, U_{i3}, U_{i4})$$

where $\overline{U}_i$ is the maximum value of the i-th user and U_{ij} is the value of the i-th user for the j-th design. Then we normalise $\overline{U}_i$ with a factor k_i so it is the same for all users, giving for n users

$$k_1\overline{U}_1 = k_2\overline{U}_2 = k_3\overline{U}_3 = \ldots = \frac{\Sigma\overline{U}_i}{n}$$

The same normalising factor is applied to all U_{ij}, and then the best design is declared to be the one having the largest total normalised value, summed over all users.

In a situation where all categories of value can be defined by curves, or where those that cannot are independent of the design configuration, it would be very useful to get a set of consensus value curves. The designer then has much more flexibility, and can determine for himself the total value of *any* specific design he wishes to try out. He can also apply probabilistic decision theory and optimisation analysis to the design process (Siddall, in press).

To illustrate, let Figure 5 represent value curves of three users for the variable weight. One reasonably equitable procedure for establishing consensus curves is proposed as follows. The mean of each curve is determined, considering only positive amounts in a feasible range. As before, a normalisation factor is determined for each user so that the sum of the means is the same for all users. The factor is used to rescale each user's curves. The curve for each design characteristic is then averaged by working with discrete intervals. This procedure breaks down when there are specification points, as in the first and second user's curves in Figure 5. A specification point is a bound beyond which the design is unacceptable, and the value at this point drops to minus infinity. In order to prevent a single user controlling such points it is desirable that no specification be permitted unless a majority of users wants one. If there then is to be one, it is taken at the mean of those specified.

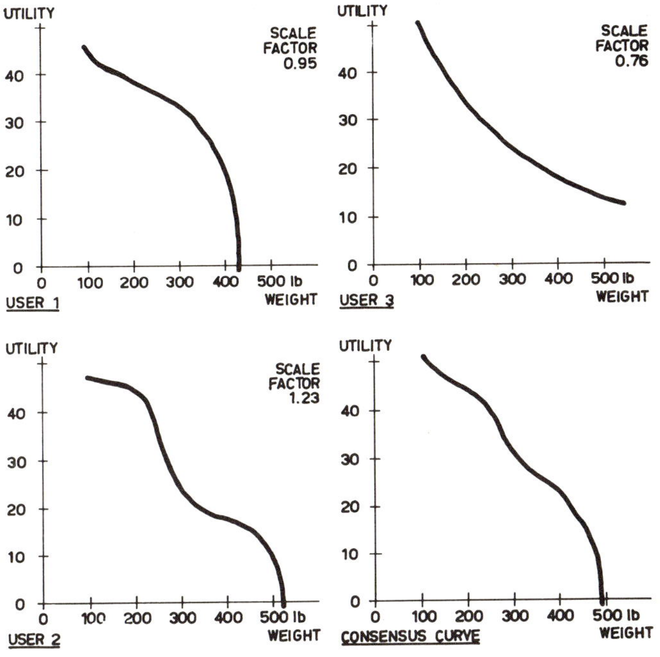

Figure 5 Obtaining a consensus curve.

USER AS DESIGNER

The user cannot completely define his values prior to the design for the following reasons.

1) He cannot fully define his values in isolation from an actual configuration because he cannot be aware of all of the ramifications of the design until it is created – for example the effect on ecology or social progress.

2) Defining values is partly intuitive and requires some immersion in the design process to be operative. The mechanism here appears to be the use of one's sense of beauty to intuitively decide if a design concept is optimum, or has maximum possible value. The idea was first suggested by Poincaré (1914) in discussing creativity applied to mathematical concepts. In more primitive times, the user was the designer. Now, if the designer finally applies aesthetic judgment as an intuitive optimisation criterion, he is acting as the user's stand in.

One possible solution to this difficulty is recycling of user input. The best design resulting from the user's value input would be submitted to the users for reconsideration of values.

CONCLUSIONS

In earlier times a craftsman designed and used his own tools. He knew from direct experience when they best satisfied his desires. From the value point of view, current design practice requires the designer to second-guess the users' desires. He must also assume that the user's value system is rational, and that the users' choice of buying one design or another is also rational. Our examination of marketing research indicates that this is not always so. Control and effective use of technology has become so important that this approach is no longer satisfactory.

Mankind has progressed so far from the primitive craftsman phase that most people have become completely disinterested in technology, and technological knowledge has become disassociated from our culture. This would seem a very unhealthy situation for a species whose whole biological evolution and essence is integrated with technology. It could be argued that our present difficulties with misuse of technology are due to this situation.

We are suggesting that one possible solution to the problem is to involve the user in the design process at least to the extent of having him specify his values rationally and directly. In order for the user to use value theory as a vehicle for participation in design he must have some understanding of the design process and engineering in general. All of this argues for more study of engineering as a general cultural subject.

STRATEGIC DESIGN FOR INTERNATIONAL HEALTH CARE

E. Matchett and K. G. Williams

Design for health care exists in a hierarchy of levels. In this it is similar to many other situations in the area of social development. In the primary foundations design is concerned with relationships between individuals, and in these the traditional, yet still essential, bias is towards those of the doctor-patient. From here it increases in complexity until design is involved in the form of the health care systems which are the modes through which medicine is integrated into the overall social structure. In between there is what might appear to be a continuous spectrum of requirements, but, on further examination, clearly defined levels of functional specification can be seen. The engineering systems within a particular health care environment; medical technology systems; and the specific requirements of community health, are typical examples.

This set of design conditions is of recent origin and cannot be traced back to the traditional practices of medicine. It is, therefore, not surprising that the sudden emergence of need has produced many strains with a concomitant, but still relative, failure to reach solutions. In turn this has produced a situation in which throughout the world the delivery of medical care, or, as it can be more formally called, health care, has reached a somewhat dangerous and critical stage (Bryant, 1969; Jones, 1970).

In the main the problems stem from the fact that in the historical development of medical care considerable emphasis was placed upon the personal relationships between the medical professions and the patients who were being treated. For the conditions then pertaining this was correct. But it created, at the same time, definite professional attitudes and, because of the authority of tradition and education, these have remained very dominant even although the social need has altered beyond recognition. With such social change and, in particular, the growing strength of science and technology, the demands upon medicine have increased enormously in scale. They have spread out from the needs of the individual to encompass much of the social and political environment.

As these circumstances developed it became impossible for the needs to be satisfied by the old structures of the medical professions and so emerged a completely new set of conditions. However, these came about virtually empirically and were not matched to the unexpected rapidity of change that has taken place within the environment they were meant to serve. So the nursing professions, all the various subgroups of medical and health care technologists, the almost infinite specialisation in medicine itself, administrators, those with financial interests, scientists, government, industry and much else started to be involved in what had now become big business as well as a significant social force.

THE MISMATCH OF NEEDS AND SOLUTIONS

This multiplicity of interests arose in and around the extensive increase in the technological and organisational needs of health care, which ranged from the use of high technology in the care of individuals to the need for a total health care system which was in keeping with the structure of any given social sector. Unfortunately, however, the development of the new needs and the emergence of the new interests were not truly related in form and time. Thus they did not match in a fundamental sense. Empiricism and pragmatism dominated the scene and, as a result, the criteria for rational design linked to change were not fulfilled.

At nearly all points sectarian interests began to dominate. The medical professions attempted to hold on to traditions and their particular status. The newer professions – nursing and then the associated paramedical groups – quickly began to see that they had to take strong action if their position in the newly developing hierarchy was to be maintained or increased. In addition, with the emergence of health care as part of social benefit within the political systems of virtually all countries, there has been an increase in what can be called the social engineering of health care; but again on many occasions the real meaning of what was required became lost in political expediency. In the end the administrators who had to attempt to resolve many of the problems on the use of available resources were left rather bemused by what was happening, and the patients, who were the real cause of it all, were often ignored as a necessary but somewhat unwanted raw material.

The relationships of these interactions are shown in outline in Figure 1.

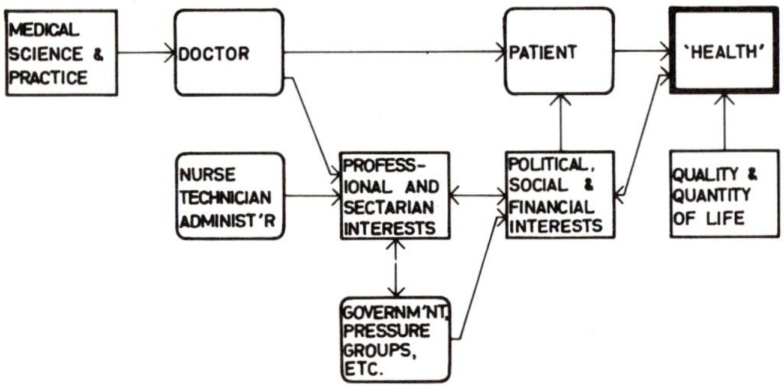

Figure 1 Some interactions producing conflicting interests in the provision of health.

UNDERSTANDING THE DILEMMA

In many ways these words may seem a somewhat harsh indictment of what has been and is the case. But it is necessary if we are to attempt to find a way out of the many dilemmas which now exist in the fields of health care.

Many factors are involved—from medical tradition and knowledge to political and social concepts. These all need to be added together so that the resources available produce the highest level of medical care for the maximum number of people. Under ideal circumstances this could be considered as a stable state with the input of all necessary ingredients and output of the desired ends.

Unfortunately this ideal case is never achieved. The various groups concerned with the overall strategy never come together in a stable form, and, in particular, the inflow of resources never matches the desired objectives.

In terms of this conference much of this lack of stability and strategy is derived from a failure in participation. Moreover, it is a failure which stretches far beyond the user and the designer because in medical and health care there are two quite fundamental questions which are not always answered, namely, "Who is the user?" and "Who is the designer?". It may seem strange that such quite basic queries should arise. But it requires only a little consideration to understand the reason. Are the users of the health care system and its facilities the medical people who care for the patients, or are they the patients who receive what should be the final benefit? The answer to that would probably vary from case to case without rhyme or reason. Similarly, when we come to the situation of the designer there would be great difficulty in defining who this is, even in the fairly straightforward case of a new hospital for a health care system. It would depend upon the 'pecking order' that has been established in the planning groups involved, so in one case the architect may take precedence and in another any member of the people involved in the project who has exerted his personality above the others.

LACK OF DESIGN PARTICIPATION

The fundamental effects of this lack of design participation are recognisable right at the beginning, when the needs of any particular part of the health care system, whether in the community or in a very special environment, are defined. Because so many people are involved it is very difficult, if not impossible, to get a true concensus of opinion upon what is required to gain the objectives of maximum benefit for the maximum number within the resources available. As a result, the chances of reaching the specifications which are so essential for adequate, let alone imaginative, design are very slight indeed.

With such a lack of real specifications it is not surprising that design for health care rarely achieves the match between supplier and user which is the essence of success. The confusion which exists at so many of the levels of health care today can be traced back to this position. And this comment applies from the design of management systems to the design of such simple matters as the hospital bed. (Note the intensive work needed by Archer *et al.* (1967) in coping with the latter problem in the context of the background we have sketched.)

APPROACHING THE PROBLEMS

To overcome this set of problems there are many interacting factors which must be brought together. At a very fundamental level there is a mismatch in communication which is worsened by the conflict in logical approaches that are applied to various aspects of health and medical care.

This stems from the attitudes and educational traditions of the many factions concerned in this wide sector of human activity. By the judicious use of logical and scientific concepts it is apparent that this lack of communications can be overcome on the theoretical plane. This approach is described by Williams (1971) and involves relating the needs of the patient to his environment and then interpreting these in a form which can be fully appreciated by the various professional groups who are concerned with maintaining such patients in balance with their surroundings.

The above interpretation is concerned with optimising the flow of resources as shown in Figure 2. The process starts with the social allocation of both economic and manpower resources and then flows to providing benefit to the patient within the particular social context. From such a general picture it is possible to move to the specific and this is so whatever the requirement, be it a single instrument, a total medical technology system, a hospital environment or a complete health care system.

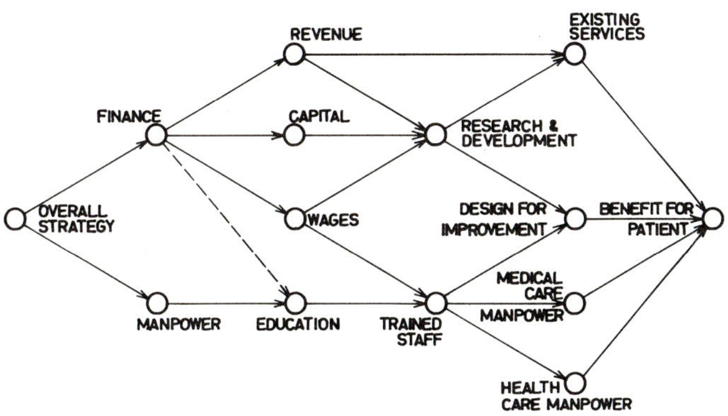

Figure 2 An outline of the necessary flow of resources in providing health care.

But such an approach clearly stays on the theoretical level unless further steps are taken to gain the full participation between all the various individuals and groups involved. Only then can a meaningful strategy for design be achieved.

Without doubt, producing this right kind of participation is the most difficult and problematic aspect of the whole subject. Yet unless it can be done there is little likelihood that we shall see a massive reduction in the major set of problems that now beset health care around the world. The right kind of participation is equally necessary whether one is concerned with resolving the ethical and moral dilemmas which can be produced by medical science and technology, with reducing the gap in health care levels between societies that have and societies that have not, or with specific design within any part of a health care system.

BLINKERED THINKING

Unfortunately, the difficulties which obstruct real participation are not simply at the level of the group, they will inevitably exist within each of the individuals who will contribute to the design. An example can be given by taking a doctor concerned with health care on quite a wide front. Such a man is shown in Figure 3. His approach to his professional existence is not and cannot be channelled into one dedicated path. It is scattered across many various personal attitudes; in this example they range from 'political' medical interests to technology and his general awareness of the world. If such channels were fairly equally balanced and could be held in view at all times this would produce relatively little difficulty. However, in so many cases one aspect of his experience temporarily becomes dominant and this then narrows his overall attitude to the larger problem.

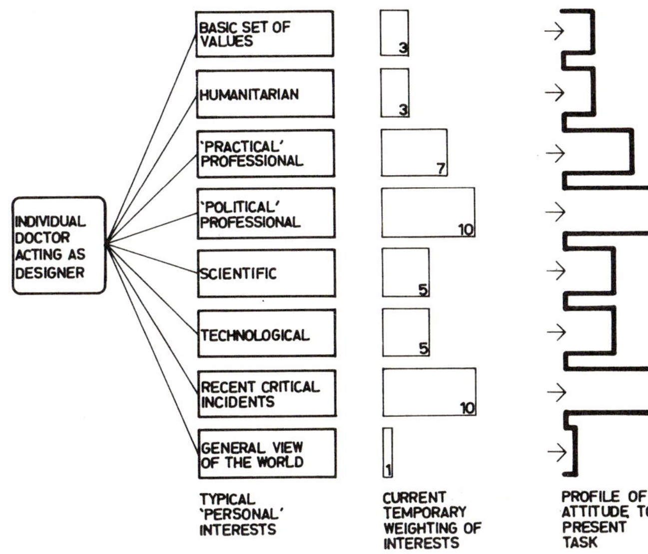

Figure 3 The confining influence on design proposals of temporary predominant interests.

So, in the particular case of Figure 3, the 'political' interests become overriding and as a result the creative capability of the doctor when he is acting as a designer becomes restricted as he tends to view his work and activity through this particular channel which has become predominant in his thinking. He is then unnecessarily limited in his approach to design, whether in the care of an individual patient or his part of a more major project in health care. Because his experiences and thinking are not all of one piece, he fails to take into account many aspects which he is capable of considering and which should be part of his behaviour in this situation. In everyday terms the interests of the moment override most of the lessons of his life.

UNRESOLVED CONFLICTS OF THE LESS INTIMATE GROUP

In the case where the design situation is related to a small number of people or to a very specific and defined technology, it is possible for such a limitation to be overcome by the normal processes of challenge and debate. However, in the case of health care many of these individual inadequacies of thinking are not traditionally so resolved due to the less intimate involvement of the very varied groups which must take part in any significant design decision.

Even if group members were all well orientated in their own specific disciplines this would cause problems. But as we have already described, in more cases than not each set of specialists will tend to be dedicated to a narrow band of active views and beliefs in his own particular area which is reduced even further by temporary predominant interests. Unfortunately there is no clear objective such as commercial success to drag the group into an effective concensus and consequently the whole tends to move forward in an unstable structure with personal dominance being more effective than rational and deeply considered concepts.

On detailed analysis many of the problems of design in health care throughout the world can be related to this simple insight. The way out, however, is not too easy to see because the processes of reaching the basic concepts and ideas upon which the optimum design should be built have not been given sufficient serious study. Consequently what we would like to call 'the strategy for international health care' which would allow the true objective of maximum benefit for the maximum number of people does not emerge. It is obscured not by a lack of good-will but by a tangle of woolly thinking.

But this situation will not be cleared through a simple systems approach built around theoretical modules which take for granted a level of human discipline which does not normally exist. The basic problem is to produce the necessary quality of such disciplines within the individual thinking and the interactions of the group. When this is done the necessary crystallisation of the optimum is more likely to happen.

Overall there must be a basic mental re-orientation of individuals and groups within the professional and organisational structures of any part of health care design. Without this, responsible participation with what is after all the ultimate user – the patient – cannot occur. And here it should be noted that responsible participation in this area is no mean achievement when it can significantly affect both the quantity and quality of human life.

We will now consider a means of reaching towards this stage.

THE PROBLEM REDEFINED

As we have noted, few, if any, of the people concerned with designing any part of a health care system will be capable, initially, of conceiving sufficiently comprehensive and meaningful thought-forms. (The term thought-form is used in preference to 'ideas' because it stresses more clearly the necessary precision and complexity.) Collectively the design group might agree upon a selection and synthesis of aspects and features of other systems of which they had direct experience, but it is unlikely that such a conglomeration would possess the quality and vital content that the new situation was actually demanding. It could be out of date even before it became a reality and it could well be riddled with serious inadequacies, ommissions and expensive errors. The problem of managing, developing and releasing the talents of the group is, therefore, primarily a problem of arranging the special kind of group involvement in which adequate thought-forms can and will be conceived.

THE NEED FOR THE EXISTENCE OF A TEAM

A team is necessary where the thought-form which is required to clothe a needs-complex is beyond the capability of the individual mind. The team should exist to produce jointly this advanced thought-form. It has to possess the special characteristic of crystallising in the mind of at least one of the group a structure that he, or she, is not able to produce without aid. It is highly desirable that this collective synthesis can be comprehended by several or all of the members of the group, but this is not essential.

Some minds more than others can readily recognise particular areas of needs. Also the clothing of certain such areas in the appropriate form does require a facility for manipulating symbols that may be beyond the vocabulary of many people. It would seem that all knowledge has to pass into consciousness translated into the symbols we have learned to apply. Hence the range of symbols, as distinct from knowledge as such, could well be the short-term limiting factor that makes a group essential for certain purposes. (The symbols of different branches of mathematics are fairly obvious examples of where particular symbols serve to release thought that would otherwise have remained unexpressed). Carl Jung drew attention to four distinctly different modes of thinking, each of which employs its own characteristic vocabularies. All four modes of thinking may be necessary to solve the complexity which exists in the problems of modern health care systems. This is another reason for employing a group for such a task and draws attention to a special problem of communication within the team.

THE MANAGEMENT OF THE GROUP FOR HEALTH CARE DESIGN

In order that a 'group-mind' shall truly come into being, the team must set out to be self-energising, self-directing and self-correcting. It must, in fact, resemble very closely the behavioural and homeostatic mechanisms of the developing human organism. Through the processes of feedback the group or organism remains clearly in balance with its specific environment and the dynamic demands of the moment.

The important question is how to achieve this natural phenomenon in a group which may well have met for the first time only a few hours before the development process begins. The group members – who will usually include people from very varied professions and disciplines – will initially be unaware of what is to be expected of them. Each will know that he, or she, has an important part to play in designing part of a health care project. But probably he imagines that this contribution amounts to giving someone else advice in respect to a particular specialist interest, or simply discussing a scheme that someone else has produced. If such attitudes are followed then it is likely that failure in co-ordination will occur.

These remarks now give some indication of the reasons behind the key features in accelerated development programmes that have been planned to implement health care projects in different parts of the world.

Here the problem is to make best use of medical knowledge and practice, health care technology and general science and technology in differing yet specific social contexts. At the same time the whole project must be viable in terms of all resources. Moreover, the people involved in design in its widest implications will be multi-disciplinary, including doctors, nurses, engineers, architects, administrators, contractors, industrial technologists and much else. It is likely that they will also come from a multi-national background.

These comments underline the size of the tasks involved and show the heights to which any 'group-mind' must climb.

ACCELERATED DEVELOPMENT PROCESS

For convenience we can divide the development process into four phases that have been shown to apply in other programmes carried out by Matchett (Matchett, 1970, 1971).

These are:
1) Preliminary Phase
2) Confrontation Phase
3) Re-orientation Phase
4) Creative Phase

We will now consider each of these in outline.

The Preliminary Phase

The correct conduct of this phase is crucial to all subsequent achievement. The members of the group come together to learn what is expected of them. Seed thoughts are sown concerning the comprehensive and fundamental nature of medical and health care and how these are and may be related to the specific project.

The idea of a 'group-mind' that is self-energising, self-directing and self-correcting is introduced. The enormity of the task and of the burden that each will be expected to carry are aired at some length. First reactions and especially fears and misgivings are dealt with very carefully and each person is given ample opportunity to opt out of the work, if he, or she, so desires, without loss of face. Some members will, for different reasons, nominate deputies or substitutes, and all who remain will make arrangements to be available for the three remaining phases.

A considerable time is given to constructing and asking questions, then having these discussed both privately and within the group. So, the first level of interaction is brought about under skilled tutorship. The group is introduced to key figures; those who will finance the undertaking, and people who have a reputation for the advanced nature of their thinking in areas of vital concern to the project.

Confrontation Phase

This is the most 'uncomfortable' phase to actually experience. With prompting and guidance from expert tutors, each member of the group is placed in a position of discovering what he or she actually believes to be the vital issues and what is truly understood about each of those. The evidence for particular beliefs, attitudes and values is sought for tenaciously and everything is challenged.

Other possible interpretations for things we know to have happened in the past are sought and there is a co-ordinated search for possibilities and opportunities that might well be worth exploring.

Key concepts such as the distinction between medical care and health care; the place of management in health services; the relationship of technology to medical practice; the use of innovation in health care development; the need to optimise the use of resources in various categories, and so on, are explored with vigour. Considerable time is given to framing, critically analysing and rephrasing definitions of terms and concepts which it is considered will play an important part in the work ahead.

Each person's ideas on objective approaches, monitoring methods and criteria, etc. are exposed, challenged and if necessary, re-defined.

The problem of how to produce the 'group-mind' is given a great deal of attention and, as well as all the other important issues, the disagreements, uncertainties and conflicts are brought out into the open for all to see and work upon.

The self-image of each of the professions represented in the group and the total equation of the design project are also brought out into the light. Where special codes of practice are likely to be working against maximum benefit for the patient (the needs of the particular project, the overall economics of medical and health care and their intelligent integration into the particular structure and so on), their justification is challenged.

A great many ways are employed to focus attention on to the objectives of the project. Particularly, the interests of the patients are brought central rather than left peripheral.

As this process continues, the scope of each person's thinking is uncovered; i.e. the number of planes that are active within his mind and breadth of content of each of these. The relevance of the active planes for the problem in hand is also considered carefully and areas of experience on which one might usefully draw are noted for subsequent action. Blindspots of knowledge and experience are also noted and some assessment of their seriousness is made. Where there are agreed critical blindspots persisting across the design team, additional knowledge might have to be brought in by adding further members to the group.

Re-orientation Phase

The re-orientation phase develops out of the confrontation phase in different ways for different groups and individuals. Occasionally it occurs very rapidly but more often it emerges imperceptibly. It might be useful here to remind ourselves of what it is the individuals or group minds are being orientated towards.

They are certainly not being programmed externally to respond in any pre-determined way. Rather they are simply becoming capable of recognising and solving the complex and unique problems which they are currently facing. These problems are unique and no one who might have a desire to manipulate the group could have prior understanding of them.

On the other hand we are not faced with a simple 'conversion' phenomenon. The individuals of the group are not drawn into a situation where they are driven by external pressure. The re-orientation comes from the growing awareness of what needs to be done and of what this implies and demands of their mental processes.

What is gained in the re-orientation process is an increase in real opportunity to perform in a certain area with true poise and professionalism. What is lost is the blinkered attitudes of extreme specialisation, the automatic responses that come from only partially digested experience and the expediency of over-simplified pragmatism.

The speed at which the needed re-orientation occurs, i.e. that which is necessary in order to be in control of the design situation, will often accelerate as the interactions within the group produce insights which are recognised to be vital. This is a self-generating acceleration where the initial gains in achievement and confidence rapidly snowball. Indeed, the process can generate enthusiasms which have to be intelligently contained by an increase in the objective monitoring procedures which the group itself generates from the foundations of technique which had previously been pointed out.

In the case where the re-orientation phase is slow to crystallise and time and money are limited, then the judgment and experience of the tutors may become important. They must discover the inhibiting factors and draw the attention of the group to them. It is important that this should not be a 'forcing function'. The seeding of a supersaturated solution to produce crystallisation is perhaps more analogous.

The tutors have another specialised role in respect to helping individuals in the group to maintain their confidence when faced by a proliferation of material which can emerge explosively and appear to be more than the individuals or the group can handle.

Some attention has been drawn to the monitoring which is a vital element in this phase. Much of this is concerned with making checks on the involvement, the objectivity and the understanding in the central areas of concern.

There are, however, two aspects which are worthy of special mention. The first concerns tests for the validity of the mirroring of the 'planes of demand' mentioned previously and the validity of their final content. The second major issue is that of finding proof for what is finally accepted as a reasonable basis for entering the creative phase. Such proofs have, of course, been sought throughout much of the preceding process, but at this critical stage they are applied with increasing emphasis.

The Creative Phase

The form of the creative phase is quite other than what might have been expected. Even although it is a group activity it more closely resembles the seemingly effortless actions of the truly creative individual designer. Once the initial phases have been successfully mastered the sheer size and complexity of the needs-complex no longer seem to be a hindrance. Ideas flow quickly and naturally and the individual mental functions appear to be almost one. It should be noted that this is not yet another manifestation of brainstorming. Here all the material is important, it already possesses major structure and it is directly related to the character of the needs-complex.

For instance, in the case of a hospital design project it ceases to be an amorphous mass of buildings, engineering elements, pieces of equipment, a computer and a rather disorganised set of various groups of people. To the 'group-mind' that has become creative it is seen as a whole. The technology from a single instrument to complex management sub-systems becomes an integrated total design dedicated to the maximum benefit for the maximum number and with best utilisation of manpower and economic resources. It will also match the many social factors in the needs-complex.

It could be felt that such a process would reduce, say, the component of architectural creativity. In fact the opposite occurs. The integration that has now taken place within the group is witnessed quite naturally and automatically. For the architect this gives release and he can now reach towards his ideal point where form fits function far more comprehensively. Similar comments could be made about many other forms of health care project design from management and information systems to complete sectors within a health care system.

Now a final word about 'the camel' – contrary to a deeply entrenched modern belief, the design produced in the way which has been described is not built up piece-meal and does not have to apologise for its deficiencies. The design is a thought-form which emerges through the controlled interactions of the group. It is a unity and it is complete overall in its important aspects prior to the design being committed to paper.

CONCLUSION

So with this necessarily very brief outline, we point out an approach to what we have called strategic design for international health care. It is clear that the description is very much a *precis*. Only the important highlights are included. Inevitably many details of management from initial formation of the team to techniques and time-tables have had to be ommitted as these vary considerably from project to project. Similarly, further possibilities such as the total integration of data-processing into the group structure have not been considered. This is an exciting and challenging opportunity but in these early days it would tend to confuse the more established concepts outlined above.

It is important to note, however, that in this description we have concentrated almost entirely on the human element in design. The specific aspects of technology itself have been set aside. This is quite deliberate. Today the stage has been reached when the technological aspects of most health care projects can be solved adequately if the resources are available. The real and major problem is how to tap this knowledge within the complexity of human activity which is involved in the overall design.

Further, it is clear that, although we have concentrated here on health care design because this is central to our present interests, the matters outlined have a much wider relevance. They are applicable to any design situation which involves multi-disciplinary design teams and specifications which take in many areas of human activity.

From the many possibilities, one could mention environmental planning, communication systems, information and management systems, major man-machine systems and any complex project with a high degree of advancing technology and scientific innovation.

It is also clear that in the team design process we have outlined there is a considerable need for participation at all levels in time and form. But to consider this merely as a reaction between user and designer would be naive to say the least. With major goal-centred activities, participation, including social as well as individual action, becomes the source of design. Without it the major mistakes, which are so well illustrated in many aspects of health care, will continue to mar the present era of human progress.

THE DISABLED USER

Stephen Platt

About eighteen months ago, in Manchester, I borrowed a wheelchair, imagined I was paralysed from the waist down and made trips around Hulme, one of the most modern redevelopment areas in Europe. I had no out-of-the-way objectives; I wanted to see if I could move independently within the area and if I could use the shops, library, schools, churches and other buildings in a wheelchair. I found it practically impossible to use any of them because of architectural barriers . . . kerbs, steps, ramps. This was doubly disturbing when one remembers that this area is the model for future redevelopment in the Manchester area.

There is a reaction in some sections of society to the mass provision of goods and services, and talk of participation by users in the design process. This is part of the general movement for change in the distribution of scarce resources and power in society. Yet to date motorcars, housing estates, social services and most other things are designed and built with little or no participation by users in the design process. Minority groups, with special needs, are penalised by this process and possibly their only solution is to form a pressure group and make a lot of noise. Marginal groups, those on the fringe of society who lack power, wealth, and often minimal resources to lead a normal life, are even worse off.

My wheelchair experiment was one of the methods I used in design research for one of these marginal groups: the disabled. Hunt (1966) points out that although the disabled suffer from the effects of design for majorities they are, for many reasons, one of the least vocal groups in society.

It is hard to define disability because the word has such a wide range of meaning and because the lay and medical terminology are often confused. Townsend (1967) gives a useful functional definition in terms of "a limitation or impairment of 'normal living' by loss, abnormality or chronic illness". He also estimates that about 1.5 million or about 3% of the population in Britain could be officially described as disabled, although he points out that studies in Denmark and Sweden, see Anderson (1964), in- n-dicate that this figure could be as high as 6%.

THE HULME REDEVELOPMENT AREA

The Human Factors Unit of the Manchester Polytechnic, under the direction of Ray Gray, has been working on the design of equipment for the disabled for some years, and the idea for this study into the living activities of disabled people began as research into the needs of the disabled and how far these needs could be met by the equipment designer. A pilot study was conducted in the Hulme Redevelopment Area in Manchester. The area was chosen firstly because it is new and should represent the most up-to-date design for the disabled, and secondly it is a clearly identifiable zone within easy reach of the Polytechnic. At the time of the study about 80% of the 4340 homes in the redevelopment area were complete and inhabited. For administrative purposes and in reality the area is separated from the rest of the city and split into three neighbourhoods by urban motorways. There is a mixture of housing type; twelve-storey and nine-storey tower blocks, four-storey slab blocks and two-storey terraces.

Ryan (1970), of the Manchester University Sociology Department, in a preliminary analysis of the data from a community research programme in Hulme, estimated the total population at the time of the disabled study to be about 11,000. The number of disabled people registered with the Welfare Department was 128 or 1.2% of the total population. However the Deputy Welfare Officer for Manchester said that he doubted whether their records covered more than one third of the disabled population in Hulme (Broadbent, 1970).

80% of the 128 disabled people on the Welfare office records were over 50 years old and 70% over 70 years old. 64% were women. There was a wide range of disability; 35% had sensal impairment (blind, partially sighted or deaf) and 69% had a wide range of physical disabilities, 21% of all disabled had some form of arthritis and 10% were paraplegic or hemiplegic (paralysed in both lower limbs or in one side of the body).

METHOD

Two methods of research were used. First, to get an idea of what it is like to be disabled, I accepted Hugh

Morris's (1970) invitation: "if you are able bodied and want to appreciate their fears . . . borrow a wheelchair and try going about your normal work". I borrowed a wheelchair, imagined I was paralysed from the waist down and made trips around Hulme. I made notes on a miniature tape recorder and later repeated the trips with a photographer. Next I went to the Welfare Department which covers this area and chose 20 people to represent the 128 disabled people living in Hulme who were on the Welfare Office reçords. I went to see each of these people with a checklist of normal living activities like dressing, cooking, shopping. 25 were indoor activities and 25 outdoor activities. See Chapin and Hightower (1966) for a fuller description of this method of investigation. Each interview took about half a day, both because of the open-ended nature of the questioning and because disabled people are starved of company and want to chat.

LIVING PATTERNS

Disabled people are virtually confined to their homes. 80% had difficulty going out and 20% never went out of their homes. Within their homes they have difficulty performing the simplest task that normal people take for granted. For example 80% mentioned difficulty eating, 50% difficulty dressing, 50% difficulty in using the toilet and 50% had difficulty washing themselves. Even the simplest task, like going to the toilet, can be an exhausting and soul destroying expedition. They have extreme difficulty doing any spare time activity – a hobby, sewing, reading or even entertaining friends. Many of the disabled spend the day looking out of the window or listening to the radio, and pass the evenings watching the television or again listening to the radio. Away from the home their lives are even more limited. The only activities performed regularly were shopping and going to a club, and the only activity mentioned as performed without difficulty was going to the hospital, because they were taken by ambulance. People could remember the last day out or holiday vividly, even though it had been years before.

VARIABLES AFFECTING DEGREE OF HANDICAP

The difficulty a disabled person has is clearly related to the severity of his physical impairment and to its constancy or irreversibility. This second point is important since two disabilities, poliomyelitis and muscular dystrophy, may have the same symptoms at a point in time but polio is a disease which causes a static disability while muscular dystrophy is a degenerative disease for which there is no cure and which causes a worsening disability.

The handicap a given disability is to normal living is dependant, however, on a number of other variables. Wright (1960) points out the importance of psychological variables, temperament and personality, in the ability of the individual to adjust to a disability. In the Hulme study, age was found to affect people's accommodation to their disability. The young people interviewed were bitter and depressed while older people tended to be more resigned and cheerful. However it is difficult, without a bigger sample, to separate the effect of age from the length of the disability, since older people tend to have been disabled longer than young people.

The family situation also has an enormous effect on the disabled person's ability to lead a healthy life. The homes of married disabled tended to look cleaner and pleasanter irrespective of whether the non-disabled spouse was male or female. Again this variable is difficult to isolate from the final critical variable, income, since married disabled tend to have more money left after all the essentials are paid, especially if the husband or wife is working. Most of the disabled living in Hulme had low incomes, whether married or single. Single disabled people received between £7.80 and £8.40 a week in sickness benefit or old age pension and supplementary benefit. Clearly an extra £10 a week might do more for their mobility and life style than any design innovation.

DESIGN CONCLUSIONS

Nevertheless, design is important both inside and outside the home. Disabled people have considerable difficulty with faulty equipment, poor architecture and thoughtless planning.

95% of disabled in the Hulme sample had difficulty preparing food and 80% difficulty eating. Disabled in wheelchairs have extreme difficulty using standard kitchen fittings in the home. The stove with a hot plate height of 33 ins. and an eye-level grill is difficult and dangerous to use for a person in a wheelchair. The sink bowl is too deep to use easily and the table too high at 36 ins. The wall storage cupboard, with shelves at 52 and 70 inches above the ground were never used. The local authority is very reluctant to alter any fittings in the kitchen.

One solution to this problem of food preparation by disabled in wheelchairs has been presented by Pugh and Associates (1969) in their design of the Drummond Centre for Adult Spastics. The working surface height is a constant 30.5 ins. (775 mm) for all fittings, and there are knee holes below the sink and hot plates.

None of the disabled with little hand control had any special implements for preparing food or eating. Agerholm (1966), in a four volume edition of *Equipment for the Disabled*, details many aids that are available for the disabled. Clearly, disabled people need more special advice and assistance from the Social Services Department of the local authority.

The disabled in wheelchairs have considerable difficulty of access to bathrooms, difficulty transferring to the bath and of transferring to the toilet. Many of the ambulant disabled also have difficulty of access to bathrooms and difficulty of balance. The Welfare Department told me it is extremely difficult to fix rails to the plaster-board walls and that they are unwilling to do any structural alterations.

The solution presented by Pugh and Associates, again at the Drummond Centre, has separated the toilet and bathroom. There is a gate hand-rail at the side of the toilet, so that the disabled in wheelchairs can transfer from the side and still have support. There is a shower fitting in the bathroom with a seat and a drainage channel covered with a grille for wheelchair access.

44% of the disabled interviewed had difficulty looking out through the window. The Jesperson building system, widely used on the site, has a structural component nine inches deep running across all windows at eye level when sitting. One disabled lady with arthritis watches the world through the reflection in her television screen because she is too low down to see directly out of the window.

All the disabled people had difficulty going out and very few went out alone. Architectural barriers begin at the door where there is a four inch drop inside and a six inch drop outside all houses and flats. Two 30 inch wooden ramps are supplied by the Direct Works Department for wheelchair users. However, they have to be put in position each time the person goes out because the door will not close over the inside ramp and the outside ramp must be taken inside, on the orders of the Welfare Department, to avoid an obstruction or the ramp being stolen.

One disabled lady had solved this problem by having a drawbridge ramp made, hinged in the centre and fixed in position. Both ramps are left outside the door. When she goes out she pulls the top ramp over into the house. She drives out over the ramps and then pulls the inside ramp back outside to close the door. In housing it is general practice to raise the ground floor twelve inches above the surrounding ground level to facilitate damp proofing. The problem has only been successfully resolved by building a

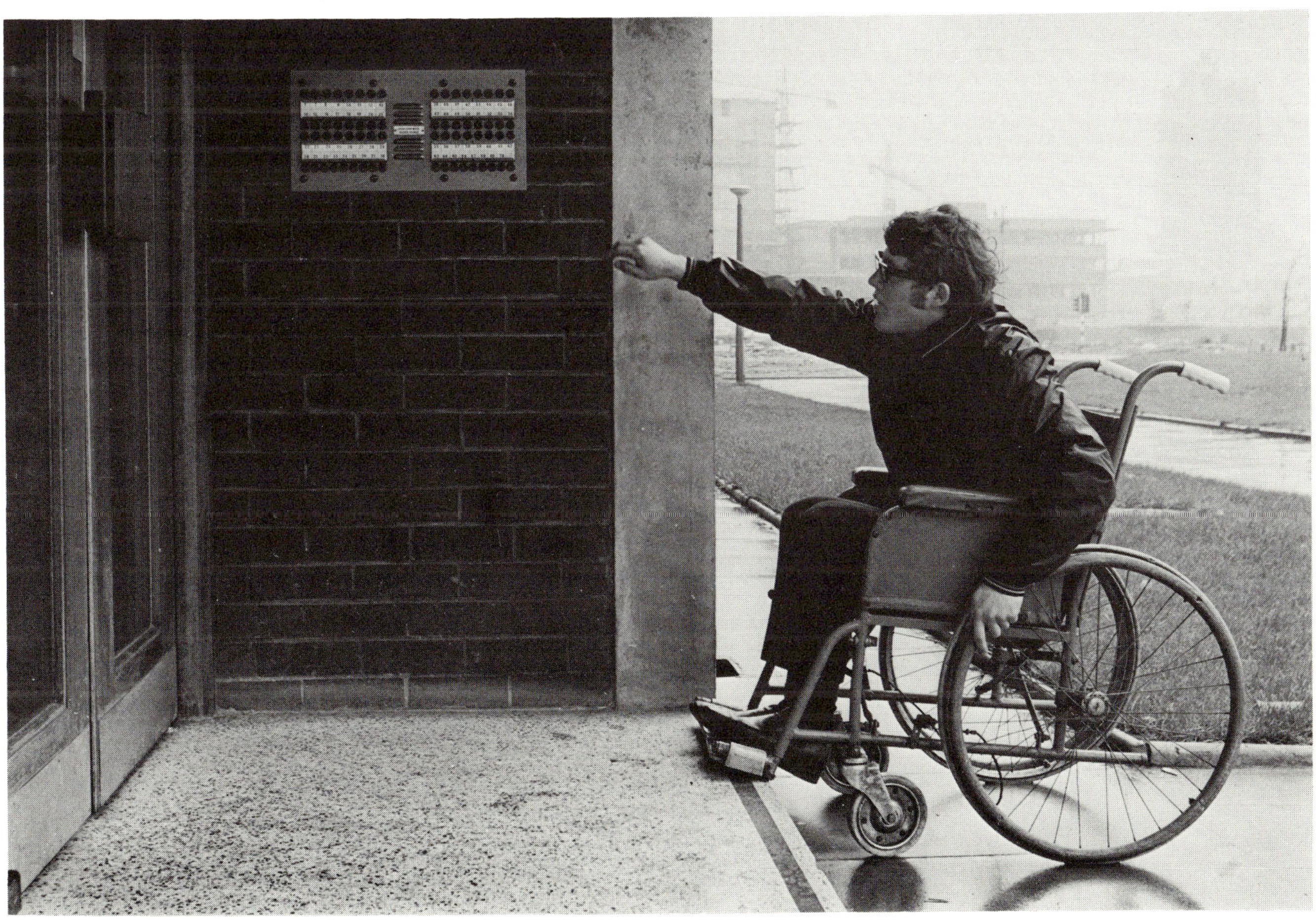

drainage channel around the building with bridges at entrances. This solution was used at the Drummond Centre. This may be an expensive solution for any but special housing, but it is difficult to justify the use of a six inch structural component across all front doors, as in Hulme, when a P.V.C. strip as illustrated in Goldsmith (1967) would serve to keep rain and draughts out. This is clearly a case of an industrial building system dictating to the architect.

Half the disabled interviewed lived above the ground floor and are highly dependent on lifts. However there is a lot of vandalism in the neighbourhood and the lifts are often broken down. This produces a feeling of insecurity in going out, especially at night.

To make short journeys within the neighbourhood, to go shopping for instance, the main problems for the wheelchair disabled are kerbs, steps into shops and buildings and the long, weather exposed journey. For the ambulant disabled person there is the long walk without sheltered resting places.

Kerbs are high in Hulme, about six inches, and there are few ramped kerbs at crossings. This can be an incredible barrier to movement for a wheelchair user, as I know from experience. Bridge (1970) in a letter to the Manchester Evening News said: "these new kerbs are going to prevent many of the old and disabled from enjoying even a minimum amount of fresh air and pleasure". There are few shops in the area and all have high steps. The library has a lift but it is at the top of a flight of twelve steps. Only in the design of the Wesleyan Chapel has some thought been given to the needs of the disabled.

Disabled people also have great difficulty making longer trips, for example to a larger shopping centre. Few disabled have cars so are forced to walk or use public transport. The neighbourhoods are bound by urban motorways with ramped underpasses for pedestrians. The Central Council for the Disabled (1969) says this type of long steep ramp is unacceptable to wheelchair users, with or without someone pushing, and to ambulant disabled. The British Standard Code of Practice (1967) recommends that where the gradient exceeds 1:20 the slope should be a maximum of 30 feet long. In Hulme the slope is 1:10 and since the floor of the underpass is 12 feet below the level of the road, the ramp is four times longer than recommended, 120 feet. In a wheelchair, I had to rest three times on the way up, clinging to the side rail to stop myself sliding back. Coming down is, if anything, even worse.

None of the disabled interviewed used buses and the few people that went out regularly used taxis. This is an extra financial burden for the disabled. Those disabled lucky enough to have an invalid car have problems in that they are not allowed to carry passengers. One lady who collects her child from school has to make her sit on the floor of the car to avoid her being seen and the consequent risk of the car being taken away from her.

CASE STUDIES

As well as these general design conclusions it has to be remembered that each disabled person is an individual with special problems and needs. To illustrate this I want to give two case studies from the Hulme sample.

Ian P. is 33. He used to be a bricklayer but as a result of an operation three years ago his left side is now paralysed. His wife and child left him and he lives off £7.30 a week Social Security and 50p for sweeping up in a local garage. He is bitter about the operation and misses the company of young people. However, he exercises every day, determined to get back control of his arm and leg. He tried working in a sheltered workshop for the disabled but could not manage the journey to work. His main needs are advice and assistance with his exercises, more money and an invalid car.

Rachel F. is aged 58 and has had multiple sclerosis for ten years. She has little control of her hands and is confined to a wheelchair. She lives with her husband, who does all the housework and nurses her. She has great difficulty eating or knitting or doing anything with her hands. She has difficulty reaching the door to let in visitors and she cannot get into the bathroom unaided. She knows the disease will get worse, but despite this she is cheerful and resigned and only regrets the burden she is putting on her husband. She needs a ground floor home and as she said "a door knocked through from the living room to the bathroom would make me more independent but I'm still waiting for council permission to have it done". She also needs the door to the balcony widening, so that she could sit out on a warm day, and a faulty double-glazed window replacing so that she could see out.

DISCUSSION

Sainsbury and Townsend (in press) in a study of the disabled in London, Essex and Middlesex conclude: disabled people, although recognising their own limitations, want to lead normal lives within society and do not want to be segregated or treated specially. This raises the question of special or general provision for the disabled. Goldsmith (1967) in a discussion between American and British Codes of Practice puts the dilemma plainly. The aim of the American

Standards Specifications (1961) is for all disabled to act normally and independently, and a national campaign was launched to make all public buildings accessible to the disabled as part of the normal user group. Goldsmith argues independence of movement without distinction is the wrong approach and that the disabled need special provision in the form of ramps, wider doors, and special toilets; and that these must be signed as available to the disabled. The cost of general provision for disabled as a part of the normal group would, he argues, be prohibitive.

Clearly both arguments are correct. It is unrealistic to rebuild all existing public buildings, and adequate provision should be provided and adequately signed in all public buildings. A *Guide to Manchester for the Disabled* (Thornley, 1969) illustrates how much work there is still to be done in this area. However, in the design of new buildings or houses general provision for the disabled can be made with a little care, at little extra cost and with a spin-off for the normal population. The half-mile exposed walk to the shops, the broken-down lifts, the long steep ramps under motorways, the high kerbs, all are an inconvenience to most people as well as being disastrous to the disabled. The bathroom where you have to push between the bath and hand basin to reach the toilet may be critical to the disabled in wheelchairs but it may also be difficult to bathe children. The bus which is impossible to enter in a wheelchair may also present problems to mothers with push-chairs and small children.

The solutions exist if we want to take them. Stepped entrances may not be necessary and cars can be made to go under or over pedestrian ways, or the transport system could be separated from pedestrians to the greater benefit of us all.

RECOMMENDATIONS

1) The local authority must try and remove some of the architectural barriers to disabled movement by putting in ramped kerbs at all crossings and by making sheltered resting places on long walks to the shops.

2) The housing management of the local authority should move disabled to ground floor flats and homes as they become available. Basic modifications should be made to these homes, perhaps for general types of disability:

	ambulant disabled	wheelchair users
with hand control	1	3
with little or no hand control	2	4

These homes could then be retained specially for disabled users.

3) Direct action in the community is needed to survey the disabled in each area and to make a case record of their needs. Then a team of volunteers should remove the unnecessary step, lower the kitchen cabinets, widen the door to the balcony, knock through the door to the toilet, take people to the shops and the club and see they get home safely, and provide company for lonely disabled people. Who will do it? Maybe the new Social Services Department will have more resources, but I think they will be just as overworked and bureaucratic as their predecessors, the Welfare Department.

4) Telephones should be provided free or at very low charges to all disabled people. Nearly all the disabled interviewed wished they had a phone, but the one person who had one was worried about paying the rental charges.

5) The design effort into new aids for the disabled must be continued and extended. There has been a lot of work on a special road vehicle for the disabled, by Ray Gray at the Manchester Polytechnic, by Bray and Cunningham (1967) in the United States and by Ballamy (1967) in Britain. There has been work on wheelchair design, again by Ray Gray at the Manchester Polytechnic, and by Udden (1969) in Sweden and at the Royal College of Art in Britain, Livesey (1969).

Many problems need to be solved. For example the disabled person with little body control but with a hoist and track fitted in his home needs a special harness to support him in an upright position so he can exercise and use the toilet independently.

6) The Architects' Department of the local authority must monitor the existing housing programmes to see that the same errors are not repeated in new housing. In this way the user would have a voice in the design process. The industrial building system which means impossible access for the disabled because of a step detail, or a structural component across the window at eye level when seated, must be changed.

7) The disabled user must be considered in the design of all new systems. For example any new transport system should allow access for the disabled. This implies that transfer points, the pedestrian way, the boarding platforms and the vehicle floor, should all be on the same level. Stops should be near people's homes, and sheltered waiting places must be provided.

POSTSCRIPT

The National Fund for Research into Crippling Diseases has awarded the Institute of Advanced Studies of the Manchester Polytechnic a grant to continue this work in depth and to interview all the disabled living in Hulme. Maybe the Social Services Department can be made to accept their responsibility for the disabled in Hulme and maybe this could produce some feedback to the Architecture and Engineering Departments to improve future development areas.

I, personally, would like to continue research into other marginal or minority groups who are also tyrannised by the design process. The old and the very young, the poor, the uneducated, the black, the migrant all have difficulties voicing their needs to designers, planners and social services departments. Is there anyone who could help me to continue this work, who knows an organisation or university department who would be interested in this research?

PLANNING AND PROTEST

John Page

It is inconceivable to me from the practical point of view, with I suppose something like 3.5×10^9 people in the world, that one should adopt no hierarchical structure for public design decision making, simply because there are too many potential participants. I think that it is also very important to recognise, and this isn't always recognised in participation discussions, that life goes on biologically, even if nothing else happens. If the time taken to make decisions becomes excessively long, then the outcome may be worse for people than making no decisions at all, or making not very good user decisions.

I think that the time scale for participation is something that needs a lot of careful thought in the design of any user participation process, because, if the time scale is infinitely long, then people's desires will only be satisfied in the infinitely distant future. By that time they will have changed biologically, cybernetically and in a number of other ways. So the design of any system for user participation must work within the framework of time, and I think that it must also work within the framework of some political structure, that enables decisions involving conflicts of interests to be made in a democratic context. I have attempted to set out our present organisational structure in Figure 1. You may replace it with alternative structures but I think there will always have to be some kind of analogous organisation to the present one, just from the point of view of practicality. In any democratic society we have the important concepts concerned with the role of the politicians. The politicians in future may operate in entirely new ways through television channels and so on, but nevertheless I believe the politician is a very important person in the public participation process for design of large systems for a number of reasons.

The politician, for one thing, has a user feedback through the political machine. He has to take account of the fact that there is a distinction between those who will want to put a tick against his name as a politician and those who will not. Most users do not have the same powers to affect designers' futures. The designer is in a stronger position to ignore his critics. The behaviour of the politician is conditioned by this feedback reaction which may remove him from the

Key to Figures

A: Administration
C: Client
D: Designer
O: Opposition party
P: Political direction
PC: Planning control
PO: Hardware production organiser
Ue: External user of other systems in the environment
Ui: Internal user of the designed system
sUe: Specialist assisting external users
sUi: Specialist assisting internal users.

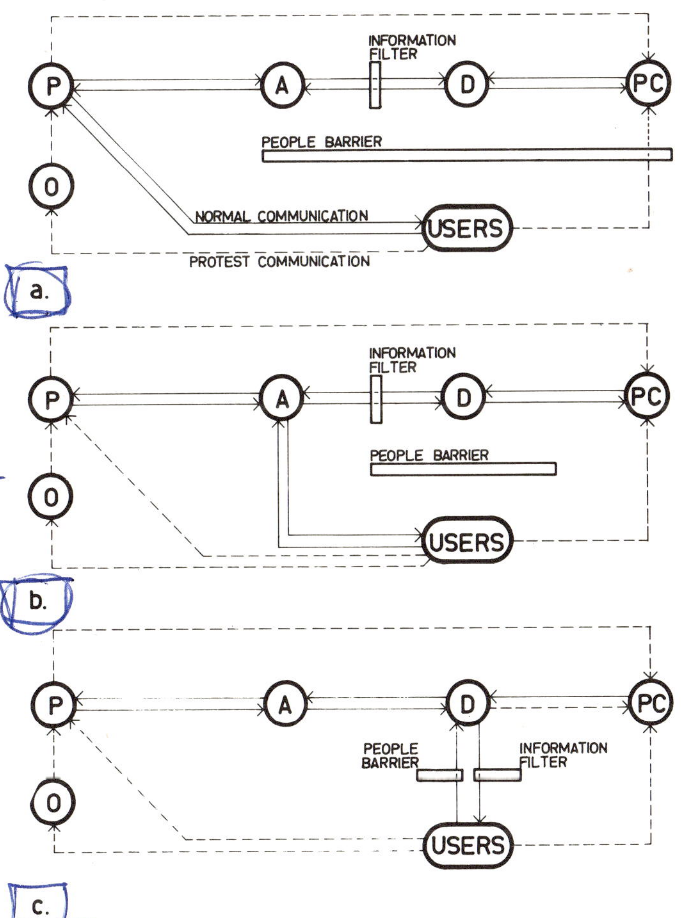

Figure 1 Design and planning organisational structure.
a) Communication through politicians only, including protest. Planning control channel effectively blocked. In totalitarian states, people barrier may extend to political field.
b) Communication through administration; protest channels through politicians, and planning control protest channel open.
c) Direct communication to designer; protest channels open through politicians and planning control. People barriers and information filters optional at choice of designer.

political scene. This has a very important influence on how the politician sees the design of large systems in relation to planning conflicts and, in particular, how he views the role of minority groups in the planning process. There is a lot of conflict to be resolved in planning decision making. The conflict is often a nuisance to the designer, and a reality to the politician. One must not assume that different types of user will agree. Design participation does not mean user harmony.

Most official organisations do not trust designers, especially as far as money is concerned. Financially, therefore, they control designers through resource controllers and other administrators of one sort or another, to make sure that the public's taxes are not dissipated by irresponsible designers. One normally has an administrative group at the centre of public control processes. At the far end there is the actual design process, where you have the designer, who may be in one of a number of occupations – he may be an architect, he may be a transportation engineer, he may be a planner and so on. Organisationally he is at the end of the chain.

Theoretically the cycle of public design is assumed to develop in a particular, logical way. Political initiation to meet public demands, administrative support followed by the allocation of design resources, then design. However it doesn't always work out that way. The designers may be shut off from the world by what I call the 'people barriers'. This is, I think, a very important psychological concept which is used in many organisations to make sure that only a moderate amount of user feedback can penetrate and disrupt action. Many Town Halls will insist that there is no direct route to the designers, and the only route from users outside into the machine is through the politicians. Alternatively, where the administration is powerful, the only route may be through the Town Clerk's office. As a user, you must communicate to the designer in that indirect way. So I think that the status quo at the moment often includes quite large people barriers between users and designers. Unfortunately these barriers are quite often erected by the designers themselves. The barriers are also sometimes erected by the administrators, because they don't want to get left out of the information chain, and they are sometimes erected by the politicians to force through a particular policy. I think that we have got to examine whether we need so many people barriers to prevent people outside getting more involved in the public design processes.

In the real world, usually there is not a precise balance between the politicians, the administrators, and the designers, and we can get three situations of power dominance. One is the design-dominated situation where the designer works himself into a very powerful position. He operates through the administration, and sometimes the politician may become a pretty small figure in the policy evolution processes. He may not be told precisely what is going on. A people filter may operate between the external world and the politician. Sometimes the channel of communication to politicians has a high input impedence and very few user signals enter the system from outside. Information tends not to flow laterally inside the organisation.

The designer, in the design-dominant situation, tends to exploit the administrator, who only passes on some of the design information to the politician, who only passes on rather less to the people. A national motorway, for instance, produces locally a kind of externally-generated design-dominant situation backed by the super system. It becomes superimposed on the local administration so that the local politician is cast in a clearly minor power role, because a motorway is a super-system concept. He has very great difficulties deciding what user communication channels should be kept open to him, because he does not feel, and actually is not, in control. User comment is therefore an embarrassment at a local level.

There is another kind of situation where the politicians become very powerful. This, of course, is an important characteristic of totalitarian states, where basically the politicians tell the public designers what to do. They operate whatever people filters are convenient between the external world and themselves to manipulate the situation. The decisions flow from politician to designers. In such situations where designers are politically dominated, people barriers may be erected by the politicians between the designers and the users, so that the designers are not effectively able to communicate with the external world for various political reasons. Finally, of course, there is the sort of classic bureaucratic solution where the administration takes over. The main product then is paper, and neither the politicians nor the designers can do very much about effective user communication in this situation. We will have to devise user participation systems that overcome these sorts of problems.

I have tried to sketch down what I thought the various participation systems were that people had discussed in this conference. These ideas are drafted in terms of rather primitive control diagrams.

Figure 2 illustrates the situation as a lot of designers used to see it. The designer lives in the world of design isolation. There are effectively two worlds – the design world and the world of users – and while the real world contains real users, the designer works with abstract users, whose characteristics he invents.

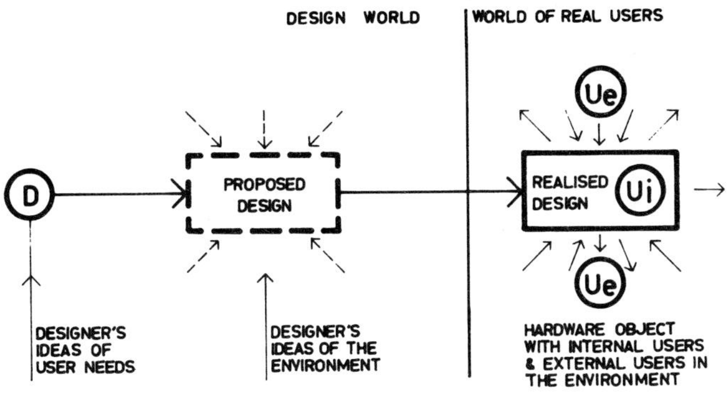

Figure 2

Eventually, when the product emerges from this "design god", it exists in the real external world. It makes an impact on the external world but not necessarily a very good one. However, because design is interactive with users, the external world acts on it. Basically the whole design set up is contrived so that users are kept out, because the process is considered an individual creative process. Only the few are trained to believe they possess the ability to do design. They will be merely disturbed by the mass of users trying to invade the design problem. The concepts used about people may be notional concepts of what the designer thought the users ought to be like, or they may involve quite refined data of a statistical kind describing how people are reacting to similar environments that have already been designed. But there is no direct questioning of users or by users in this kind of design arrangement. This conference is about abolishing that particular system.

A slightly more refined system, illustrated in Figure 3, is one where the designer is working in relation to some sort of official design control framework, say a planning control department. He produces a plan, which is shown in the diagram with a conceptual environment round it. This is the environment in which the designer believes he is designing. This design is regulated through some planning control process, for example national rules are conveyed through to the administrative machine concerning

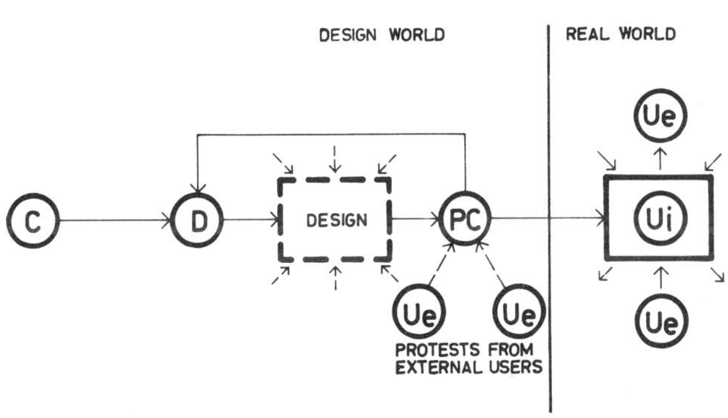

Figure 3

planning to control the designer. The consequent design is monitored by legal rules which preserve to a greater or lesser extent – and often to a lesser extent – the rights of the citizens whose activities would be affected by the new design. The actual design, of course, may take place within a concept of the external environment of the design which may bear no relation to the actual environment. The actual environment of the design is caught up in a technological spiral of change. It is dynamic, changing all the time. The interactions between the designed object and its environment are consequently shifting also all the time. The planning process still leaves the external user protected only by an indirect agent acting in his interests, namely the Planning Control Officer, who is standing in to represent user interests. As we know very well the future environment of towns is out of control, the protection apparently offered the user by the planning machine tends to be undermined by the events that follow. The outcome is really what we are getting at the moment environmentally in towns, e.g. a design situation out of environmental control.

The next stage of complexity is illustrated in Figure 4. The user has reached the role of an initial information supplier. He has the opportunity to tell the designer,

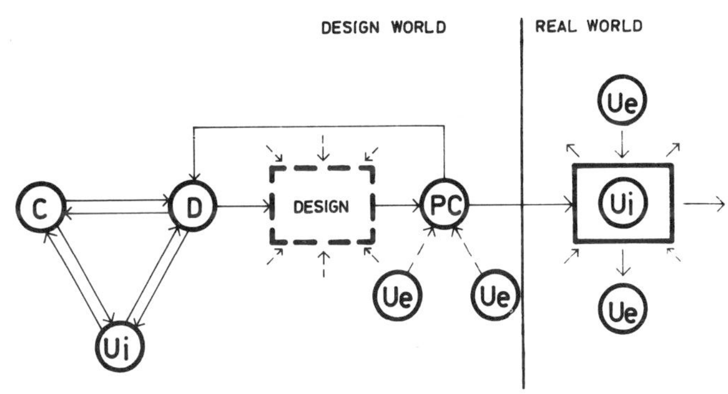

Figure 4

as an external user, at the start of the process, something about the design situation as he sees it. Thereafter the designer proceeds in the conventional way. This is a pretty primitive level of participation, as far as I can see, because the user only sets the basic performance requirements, and does not interact any further.

The next situation illustrated in Figure 5 is what I call retrospective feedback. The client tells the designer what he wants. A plan is produced which is processed through the planning control machine. This eventually produces an object in the real world with users inside it. Then there can be a feedback to the designers by dissatisfied or satisfied users within the building and dissatisfied people in the environment. The dissatisfied in the environment can usually only feedback through the planning machine, through the Law, or through

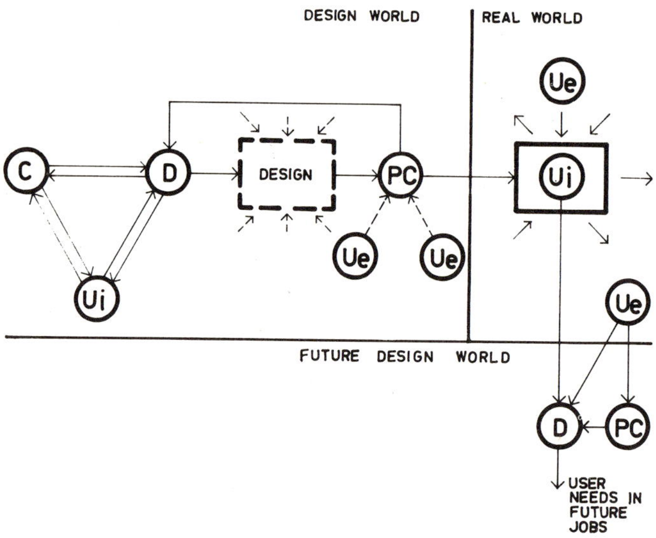

Figure 5 Retrospective user participation for feedforward to future designs.

some other government channel. The users can feedback through their own organisation. Such information really tells designers what they have designed incorrectly for users. Thus, when they design again, they have the opportunity to take certain user viewpoints about previous dissatisfactions into account. Design by retrospective feedback doesn't help the existing design. All it does is to help the receptive designers improve their next designs. However, one can get a long term design improvement by retrospective feedback. The level of participation in a sense is an accidental one, based on adverse reactions to a design and not on positive user participation in the design process as an anticipatory procedure.

One can go to the next stage of sophistication, as Figure 6 shows. Here you have a client linked with a designer who produces the design. The designer always has to submit this design to the planning authority operating within the national rule book giving guidelines on what is allowed to happen, but, in this case, the design is actually shown to internal users before construction as well. They then have the opportunity of making comments to the designer,

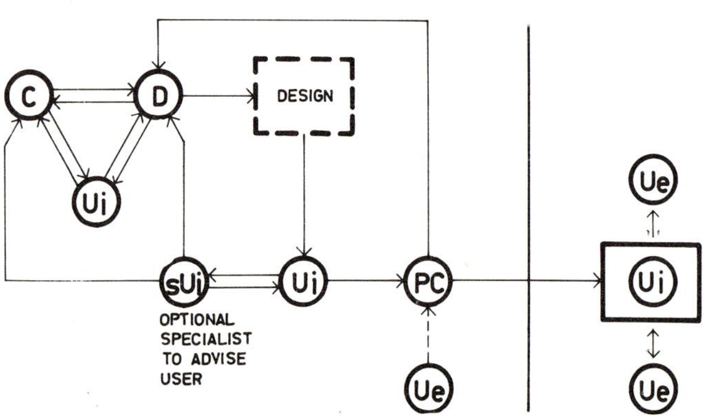

Figure 6 User based design criticism.

who can then reconsider his design and produce alternative proposals until either all internal users are satisfied or the degree of user conflict is minimised. The user criticism that takes place is not criticism of the object but criticism of the plans which are going to produce that object. Here the user has moved a little bit more towards the initiation stages of design, and is out of the post mortem situation. I suppose that, in terms of user participation, this is a big advance compared with the earlier schemes. It is the situation that we are getting at the moment with the more progressive designers but it is not a very advanced system, I think, in terms of what we have heard in this conference.

The next stage of sophistication, shown in Figure 7, is for the client to agree to let representative users in his organisation work together with him as a team, thus instructing together the designer who still is behaving in this process as a conventional designer, in the sense that all the design expertise rests with him. He is a design god still, but now only a demigod, because he gets kicked by some of the angels.

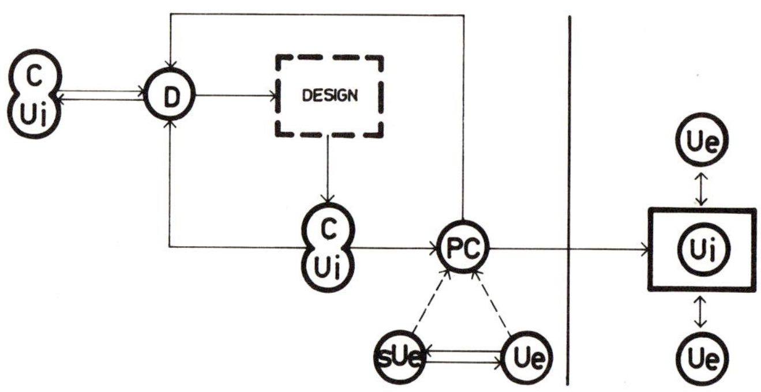

Figure 7 Client/user collaborative design – external users using professional expertise.

(This may not be allowed in heaven but we live here on earth.) The design is produced. It is then shown to the client and the users as a team who can comment further. The design may or may not be shown to external users and very often is not. You will notice that, with this design process, the internal users are included at the beginning of the process, but the external users are still in no better position to comment. It is an accident of good social sense if external users are shown the design at this stage. They can comment, however, directly to the planning control administration, which is required to hear their case. Alternatively they can comment back directly to the designer, who is not required to hear their case. This can only be done if plans are published in advance of construction. Eventually the designed object emerges in the outside world. I call this process joint client/user instruction of the designer.

You can introduce into this process a design specialist operating in a contrasting role to his conventional design specialist role. The actual users may be unskilled in the appraisal of certain aspects of the design and its environment. For example, if you are a user located in the external environment of a new scheme, you may find a planner or an architect who is perhaps living in the same area which is about to be bulldozed down, say, to make a motorway. You can then take the professional man into your user team as an expert adviser to your group, so that you reinforce the feedback channel to the planning control machine with a good deal of professional design expertise. This development of professional expertise in the consumer reaction channels is something that is beginning to happen now, i.e. the evolution of the professional anti-designer. I could envisage the Trade Union movement, for example, demanding a policy that required that all plans, say, of new factory buildings, affecting workers in an organisation, should be shown automatically to representatives of Trade Unions acting in a user technological assessment role. These representatives could have professional design support, as user design critics who could advise the Trade Union officials on the faults of the design. Such people could well be fully qualified professional designers to advise the users on how best to feedback their views to the designers. This situation does give the possibility of greater protection of the user interest by specialists operating on the consumer side. It raises some interesting problems of professional design ethics. The designer, however, is still in the design seat, and we still have not reached user design at this stage.

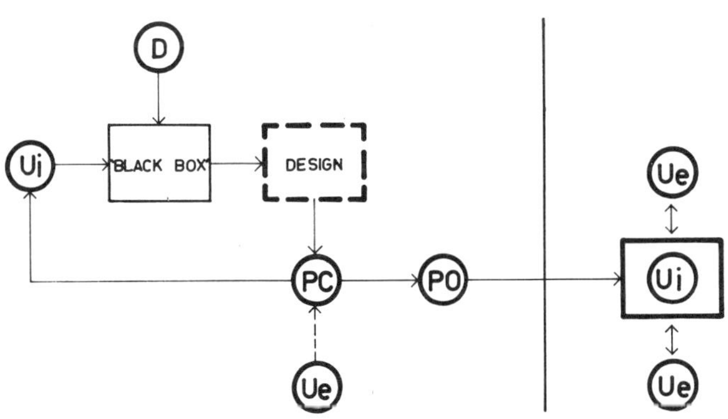

Figure 8 The user as a technologically assisted designer – the designer as a rule book writer for design black box.

The next stage of sophistication, illustrated in Figure 8, is the concept of a design 'black box', in which the need to know the technical rules of design is removed by a technological design rule-book writer who produces a design black box that makes the professional design rules freely available to anyone, using computer techniques. The box might have the structural design codes and so on written into it. Now the user himself can produce a plan, which is technologically competent. This he must feed through the external planning machine which exercises the social controls. Thus, we can envisage a system of user design where the function of the professional designer is to write the rulebooks which are fed into design black boxes, which then enable the user himself to produce competent designs which are built in the real world to match his needs. Whether you would show these plans to people in the external environment or not, would be for the user designers to decide. In interactive situations one always has to plan for two sets of users – the users of the object and the users of the environment of the object. Unfortunately many plans for user participation only deal with one of these concepts, usually the users of the object rather than the users located in the environment of the object.

Clearly this is a much more sophisticated level of participation than the earlier concepts listed. I think a number of papers to the conference have suggested that this is perhaps the process that we should be moving towards. In fact it is possible to modify this process further, and put the planning regulations into the black box. So we could eliminate the planning feedback channel, as an external channel, if the planning rules were defined. However, if we did that, then we would eliminate the external feedback channel from the general public – which might be undesirable. Automatic checks on designs in the design black box to ensure they meet the planning requirements, is a long term possibility. This idea is implicit in several of the communication systems discussed during this conference.

The next field that we discussed was user participation through simulation situations. This process is illustrated in Figure 9. In this situation, as far as I can see, the designer is still in charge. The user and client tells the designer what he wants, who then produces design simulations which are then experienced by the users, who can then comment. We are thus back onto a closed design cycle where the user is not designing but is commenting on the design through some simulation of the designed object. When the simulation is considered acceptable, the

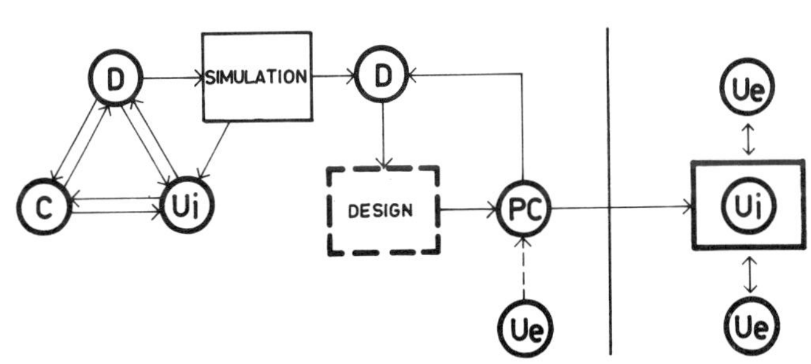

Figure 9 Designer controlled simulation for identifying user input objectives.

simulated design is then translated by the designer into a workable design. It is then processed through the planning machine, and, if agreed, the designer produces the object in the real world. There is the discretion to allow the external world to intervene in the design process using simulation, allowing them to experience what the environment of a design is going to be like. For example, you could simulate the noise from a motorway at different distances and let people likely to be situated in the external environment experience this. One would find out then what they would tell the planners about the success or failure of public noise control processes. I think that what they would tell the planners at the moment would be pretty frightening. Alternatively, simulation can be purely simulation for the benefit of internal users.

The next modification is shown in Figure 10 where the user himself operates the simulator and chooses between alternative designs by simulation. The simulation then has to be translated into a design for a real object. The translation from simulation to the real object is done by the designer as a hardware exercise lying between user interactive simulation and the production of the plans for the construction of the object.

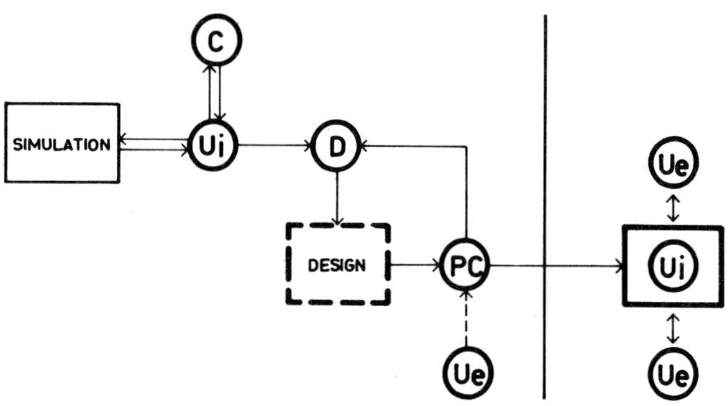

Figure 10 User controlled simulation for identifying user input objectives.

The next concept of user participation is more sophisticated, as Figure 11 shows. This provides a situation where the users and the people in the environment of the potential object collaborate in design – this is what I call interactive social design. They both operate together on the black box and they both receive the simulations. They have a joint feedback simultaneously. They can both operate together in an interactive user design dialogue. The designer can enter the process in terms of technologically writing the black box rules, the planners can enter in terms of writing the necessary planning constraints on the black box rules. A simulation is produced which is actually experienced by both sides. After agreement has been reached, the drawings, etc., can then be produced by the designers as just a technical matter, and the object can then be produced.

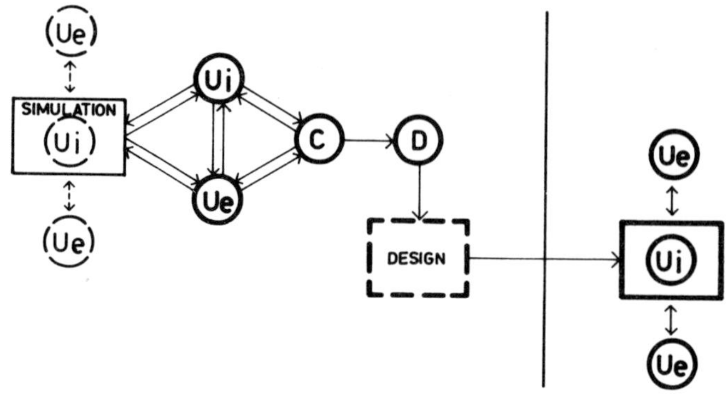

Figure 11 Collaborative design between external and internal users, using simulation to produce agreed user input objectives. (Note absence of planning control.)

Finally we have the possibilities of user-controlled adaptive design, in which the user, through some environment-modifying black box, modifies the environment in which he is situated, thus producing a feedback to himself. In the adaptive environment, the user experiences the environment which he has designed, and then modifies the design to suit himself. Eventually we come to the user as controller being totally replaced by the environmental black

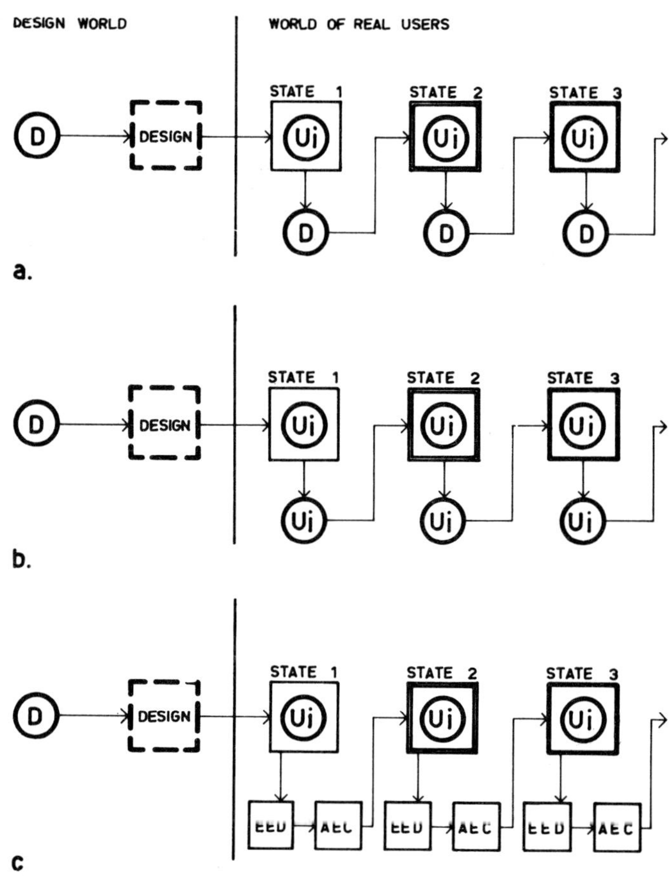

Figure 12 Adaptive design systems.
a) Designer controlled adaptive design.
b) User controlled adaptive design.
c) Automated adaptive system based on an environmental need error detector (EED) and automatic environmental controller (AEC).

box controller, which designs the environment for the user by feeding back the users responses to the environment and sensing whether he is satisfied with this change. In this sense the black box has become the designer (except that it needs a black box rule writer). This is the far end of the user participation scale – unless you go right off the human participation scale, in which case you come to design machines designing environments for machines feeding back to the design machines. The user is then eliminated. The final point reached is that of artificial intelligence, as one has created self-reproductive machines that design their own environments, so that human users no longer need to exist, and they are irrelevant to participation.

I think my comments summarise the breadth of spectrum of user participation within which we are likely to work. Clearly we have got lots of possibilities of advances in the wide field of user participation in design. Some of the ideas are clearly a long way off in terms of practical feasibility. The black box interactive system is not going to be a thing that everybody is going to set up tomorrow, I suspect. It might be useful therefore, if we were to discuss improved user participation in design on three time scales: 1) what might happen now, 2) what might happen say ten years ahead, 3) what might happen by the year 2000 A.D. We could, from the ideas of this conference, perhaps identify certain fields where effective application of new ideas is possible now, certain other fields where there is considerable preparatory work to be done, but where one can see some application within a fairly short period of time. Finally we might review the long term scenarios for user participation, requiring a lot more detailed development.

CLOSING COMMENTS

J. Christopher Jones One of the nice things about this conference is that it has brought a mixing of political and social questions with people from the arts – a lifting of art and design into the area of life rather than hardware, into living patterns and activities rather than utensils. This is a new marriage; a new media mix, in McLuhan's terms. "Design Participation" is clearly a new mix of things which have previously been kept separate – not since the beginning of time, perhaps, but certainly since the Renaissance, certainly since Descartes made his fearful cut between mind and body, or was credited with so doing. If a new mix occurs of things which have been separated for a long time, it is important, first of all, to have a pause, so that we can adjust, each to each other.

We have heard a lot of hints about how design has to change before it is fit to enter the social realm. But not only is there something 'un-fit' about the design realm when it is lifted from hardware to software, there is equally something very 'un-fit' about the socio-political realm – it is not 'fit' to be in the presence of what is good about things creative. There is something very un-couth about politics and the social order we have, which is not present in things to do with the arts. Society, too, will have to undergo a transformation before we can expect much of a mix, let alone new possibilities to emerge.

Another remark I would like to make about this conference, is that it has presented twin mirrors. It has presented a mirror for paternalists, who are the designers, often the worst paternalists, and also one for the socialists. We heard from Jeff Nuttall that there is something utterly deadly about the words 'function', 'need', etc. He said that to define a need is to define a person, and he implied that as soon as the interface between an individual and the rest of society is defined, that person's potential is fixed and therefore lost. That is the worst of things to do; that is paternalism in action. To define functions and needs is totalitarian.

The necessary re-orientation will depend in important respects, I am sure, on what Yona Friedman is writing and doing. The idea of the wholly neutral infrastructure, which does not invade the person at all, but leaves him full of possible opportunities (and also helpless in the face of all the other people), depends on trust and personal risk, rather than on law and order imposed by designers or policemen.

A nice phrase which Robin Roy brought to our notice, was "fostering vivid perceptions". This is one of the good things which the socio-political realm has to learn from the design and artistic realm – the ability to have vivid perceptions. The vivid perception of the future, of possible change, is so lacking in the socio-political realm. "It can't be done" is the standard response in politics, but it is not the standard response in art and design, where people are prepared to try out seemingly crazy ideas because they may lead to good ones.

The message in the mirror for the socio-political side of ourselves, is that the person is potentially bigger than the society of which he or she is a member. Also, and for note by the mathematical modellers in particular, the greatest and most important part of the person (that which we protest most about when it is threatened) is that part which is not modellable. It is not capable of being built into an equation because even the person to whom it belongs is not wholly aware of it. It is unique to the experience of the individual.

But, in our professional roles, we are transparent and frighteningly simple. We heard from Chris Evans that doctors, when they are diagnosing ulcers, employ very simple algorithms. Yet they take seven or more years of training to acquire these simple algorithms – along with something else, which one wonders about. Every professional role shrinks us into being simple algorithms. If we can move away from being professionals, we will be moving into personal growth, rather than personal shrinkage.

Robert Jungk I have missed, until this final session of the conference, and with the exception of Markus' paper, any discussion of politics. Having come from the Continent, where, whatever you discuss, from the size of a chair to the weather, societal implications inevitably arise (you can be sure that some Marxist will tell you why it is that way and why it is so bad), I at first experienced a sense of relief not to hear these arguments. But after two days, I became nervous, like someone who is used to noise in his work, and who then goes on holiday and does not hear the noise, who says, "But the world is full of noise, what is happening?" So I was glad that eventually we got around to bringing in some of the noise of politics.

I would like also to comment on something else which I felt was missing. We do not perceive deeply enough, I think, the possibility and the necessity for a very big change in man. Some people say it is impossible to have a new man unless we have a revolution. I feel this may be right, in a way, but it is certainly not entirely right.

Let me relate my own experiences in the past seven days. I came here from another conference, which was sponsored by a Swiss pharmaceutical company which felt that they had a bad image in society, that selling pills was not enough. So they called a conference on "Biology and Ethics". They had Nobel prize-winners, etc., there, mainly people between fifty and seventy years old. Now, coming here and meeting the people here on the very first day, I saw that this was not just a younger crowd, but a one hundred per cent different crowd. I feel that we have a lucky break in continuity between generations, which is much bigger than we realise, and which means that even without a revolution, we have had some kind of deep change in people.

What I missed, therefore, was discussion of what kind of changes in people might occur in the next twenty or thirty years. When we talk about participation, we have been talking about the people who are in control today. If we could have a new Karl Marx who would base his vision of a new social order not only on economic change, but also on psychological or anthropological change, then I think we might be getting somewhere.

What I would like to advocate is that people in the design world should begin to look deeper not only into the political scene, but also into the possibility of helping people to change. Participation may be one educational approach towards this. Most people who talk with citizen groups about participation are appalled by their lack of knowledge. But I wonder if this lack of knowledge may not, in a way, be an advantage. Knowledge is the frozen things of yesterday – all the books are tombstones on the course of our history. People with a lack of knowledge may be able to look at things in a more original, more creative way. So I think that people who are experts should go to the people who have a lack of knowledge, and say "We can help you with our knowledge, but you can help us by the way you see things when your view is not clouded by all our knowledge of feasibility, of procedures, and so on."

I have had a little experience of doing that, and it is clear that it is too much to expect the people to respond to you right away. After all, they have been deprived all their lives of any encouragement to their imaginative abilities. What one has to do, is to establish an atmosphere of confidence and trust, in which you can honestly say to people, "Your ideas may be better than mine." If you can establish that kind of co-operation, you establish a process where no longer do you listen and say "What can I do for him," but ask yourself "What can he do for me?" In that way, you may receive an invaluable challenge to your abilities, which will force you to re-think all that you have been doing before.

But the question remains open as to how far this process can go, how far we can open people up, and how far we can undo so many years of conditioning. Is it possible at all? I have found that it is possible, and, especially, Danilo Dolci, working with peasants in Sicily, has found that it is possible, to a certain level. To go further than is possible at an amateur level, perhaps we need the help of psychologists and anthropologists.

How can we undo generations of conditioning? Certainly, a real start can only be made in the schools. I believe that we do not do nearly enough towards getting new school systems that are geared to letting young people develop their own creative potential. But we can't wait. We have to work now with their parents, too.

In the next thirty years, as we meet mounting crises, the danger is that there will be a very strong argument that we do not have the time for participation. The people themselves may turn to the technocrats and say, "This is too difficult for us, we have to act fast, so do it for us." There is a great danger that we will short-circuit the decision process, and hand it all over to the technocrats.

There are two responses to that, I think. One is that, until the crises become really unbearable, we should try to go on with the participation process. In ten years, it may not be possible to participate, because the political structures will have become so firm, and so repressive. If we start now, perhaps we shall create a sort of underground of people who will survive the technocratic period. The second response is that we could talk not about participation at the moment of decision but about participation at the moment of idea generation. I think that it is important to get as many ideas as possible – we now have too few ideas. If people get used to contributing ideas to society, then this would make the decision process richer and more varied, it would be possible to draw from a larger pool of possible conceptions.

Now, I am convinced that the current economic depression, in which people are out of work or on short time, terrible though that may be, offers the opportunity for people to have the time to occupy themselves with such new things. We should use the idle time which is forced upon us by economic depression, to preparing people for doing better next time.

I am not saying that we don't have to have a radical change, but we can begin the preparation for this radical change. As a prognostician, I don't think this change will take place before the end of the century. We will have to suffer first from the lack of foresight of our fathers and grandfathers. After that, something radically different can come, but it won't come on its own; it has to be prepared.

REFERENCES

ABRAMS, M. (1971), Mass Views of the Future: a report from the Netherlands, *Futures, 3*, 2.
ACKOFF, R. L. (1969), Lecture Notes; *Principles of Planning*, University of Sussex, January.
ACKOFF, R. L. (1970), A Black Ghetto's Research on a University, *Paper presented to the Operational Research Society of America, Philadelphia Chapter.*
AGERHOLM, M. (ed.) (1966), *Equipment for the Disabled* (National Fund for Research into Crippling Diseases).
AMERICAN STANDARD SPECIFICATION (1961), *Making Buildings and Facilities Accessible to, and Usable by, the Physically Handicapped* (American Standards Association).
ANDERSON, B. R. (1964), Psysisk handicappede i Danmark, *Socialforskningsinstittutets, Publikationer 16,* Copenhagen.
ARCHER, B. *et al.* (1967), Design of Hospital Bedsteads, *King's Fund Report.*
ARCHITECTURE MACHINE GROUP (1971), Computer Aids to Participatory Architecture, *National Science Foundation Proposal* (Architecture Machine Group, MIT).
ARNSTEIN; S. R. (1969), A Ladder of Citizen Participation, *Journal of the American Institute of Planners, 35*, 4.
ASHBY, W. R. (1956), *Introduction to Cybernetics* (Chapman and Hall).
BALLAMY, D. M. (1967), *Personal Communication to Stephen Platt.*
BERRILL, N. J. (1955), *Man's Emerging Mind* (Dodd Mead).
BLISS, P. (ed.) (1967), *Marketing and the Behavioural Sciences; Selected Readings* (2nd edn., Allyn and Bacon).
BRAY, P. and D. M. CUNNINGHAM (1967), Vehicles for the Severely Disabled, *Rehabilitation Literature, 28*, 4.
BRIDGE, E. (1970), Why are Kerbs so High? *Letter to the Manchester Evening News,* 13 June.
BRITISH STANDARD CODE OF PRACTICE (1967), *CP96: Access for the Disabled to Buildings, Part 1* (British Standards Institute).
BROADBENT, P. (1970), *Personal Communication to Stephen Platt.*
BRODEY, W. (1967), Soft Architecture: the design of intelligent environments, *Landscape, 17*, 1.
BRODEY, W. and A. JOHNSON (1970), Dialogue and Exploration of Context: properties of an adequate interface, *Proceedings 4th Annual Symposium, American Society of Cybernetics.*
BRYANT, J. (1969), *Health and the Developing World* (Cornell University Press).
BUILDING PERFORMANCE RESEARCH UNIT (1967), Measurement and Appraisal of Building Performance, *Architects' Journal,* 20 December.
BURNSKILL, R. (1970), *Vernacular Architecture* (Faber).
CENTRAL COUNCIL FOR THE DISABLED (1969), *Planning for the Disabled in the Urban Environment* (Central Council for the Disabled).
CHAPIN, F. S. and H. C. HIGHTOWER (1966), *Household Activity Systems: a pilot investigation* (University of North Carolina).
DIAMOND, S., R. S. BALVIN and F. R. DIAMOND (1963), *Inhibition and Choice: a neurobehavioural approach to problems of plasticity in behaviour* (Harper and Row).
DONNISON, D. (1970), Government and Governed, *New Society*, 5 March.
DUKE, R. D. (1964), Gaming Simulation in Urban Research, *Institute for Community Development Services paper* (Michigan State University).
ELLUL, J. (1954), *The Technological Society* (Random House).
ENVIRONMETRICS (1971), *City* (Environmetrics, Washington, DC).
ENZER, E. *et al.* (1969), A Simulation Game for the Study of State Policies, *Report No. R9* (Institute for the Future, Middletown, Connecticut.)
FELDT, A. G. (1968), Community Land Use Game, *Center for Housing and Environmental Studies paper* (Cornell University).
FEO, A. (1970), Operational Games: their application to socio-technical problems, *M.Sc. Dissertation,* University of Manchester Institute of Science and Technology.
FORRESTER, J. W. (1961), *Industrial Dynamics* (MIT).
FORRESTER, J. W. (1969), *Urban Dynamics* (MIT).
FORRESTER, J. W. (1971), *World Dynamics* (Wright-Allen).
GOLDSMITH, S. (1967a), *Designing for the Disabled* (RIBA).
GOLDSMITH, S. (1967b), The Disabled: a mistaken policy? *RIBA Journal,* September.
HOROWITZ, I. (1965), *An Introduction to Quantitative Business Analysis* (McGraw-Hill).
HUNT, P. (1966), *Stigma: the experience of disability* (Chapman and Hall).
JONES, I. M. (ed.) (1970), *Health Services Financing* (BMA).
JONES, J. C. (1967), Trying to Design the Future, *Design,* No. 225.
JUNGK, R. (1969), Technological Forecasting as a Tool of Social Strategy, in ARNFIELD, R. V. (ed.), *Technological Forecasting* (Edinburgh University Press).
KELLY, G. A. (1955), *The Psychology of Personal Constructs*, Vols. I, II (Norton).
KEYES, L. C. (1969), *The Rehabilitation Planning Game* (MIT).
LASSWELL, H. D. (1959), Strategies of Inquiry: the use of observation, in LERNER, D. (ed.), *The Human Meaning of the Social Sciences* (World Publishing Co.).
LAVIN, M. (1971), The General Recognizing Evolutionary Entrance Tracker, *M.Sc. Thesis,* Massachusetts Institute of Technology.
LEONHARD, D. (1967), *The Human Equation in Marketing Research* (American Management Association).
LEVIN, P. (1971), Participation: the planners v. the public? *New Society,* 24 June.
LIFSON, M. W. (1962), Criteria and Value Systems in Design, *Proceedings Second Conference on Engineering Design Education*, University of California at Los Angeles.

LIVESEY, R. (1969), The Royal College Wheelchair, *Engineering*, October.
LORENZ, K. (1966), *On Agression* (Harcourt, Brace and World).
McCARTHY, J. (1966), Information, *Scientific American*, *215*, 3.
MACKENZIE, W. J. M. (1967), *Politics and Social Science* (Penguin).
MARKUS, T. A. (1967), The Role of Building Performance Measurement and Appraisal in Design Method, *Architects' Journal*, 20 December.
MATCHETT, E. (1970), Idea Development and Advancement in Engineering, *Systematics*, *8*, 3.
MATCHETT, E. (1971a), Techniques of Total Development, *Systematics*, *9*, 2.
MATCHETT, E. (1971b), Extending the Designer's Mental Powers, *Electronics and Power*, February.
MAVER, T. W. (1971), PACE 1: an on-line design facility, *Architects' Journal*, 28 July.
MINISTRY OF HOUSING AND LOCAL GOVERNMENT (1969), *People and Planning, Report of the Committee on Public Participation in Planning* (The Skeffington Report) (HMSO).
MITCHELL, W. J. (1970a), Computer-aided Spatial Synthesis, *Proceedings of the Fifth ACM Annual Urban Symposium*, New York.
MITCHELL, W. J. (1970b), A Computer-aided Approach to Complex Building Layout Problems, *Proceedings of the Second EDRA Conference*, Pittsburgh.
MORRIS, D. (1969), *The Naked Ape* (Bantam).
MORRIS, H. (1970), The User's Point of View, in ROYAL INSTITUTE OF BRITISH ARCHITECTS *Planning for the Disabled in The Urban Environment* (RIBA).
NEGROPONTE, N. (1970), *The Architecture Machine* (MIT).
NEGROPONTE, N. and L. GROISSER (1970), The Semantics of Architecture Machines, *Architectural Design*, Sept., and *Architectural Forum*, October.
von NEUMANN, J. and O. MORGERNSTERN (1944), *Theory of Games and Economic Behaviour* (Princeton University Press).
NEWELL, A. (1966), On the Analysis of Human Problem Solving Protocols, *Proceedings of the International Symposium on Mathematical and Computational Methods in the Social Sciences*, Rome.
NICOSIA, F. M. (1966), *Consumer Decision Processes; Marketing and Advertising Implications* (Prentice-Hall).
OGBURN, W. F. and D. THOMAS (1922), Are Inventions Inevitable? *Political Science Quarterly*, *37*, p. 83.
PACKARD, V. (1958), *The Hidden Persuaders* (Penguin).
PARKER, D. H. (1957), *Philosophy of Value* (University of Michigan Press).
PERRY, R. B. (1954), *Realms of Value* (Harvard University Press).
PHILLIPS, C. F. (1968), *Marketing Principles and Methods* (Irwin).
POINCARE, H. (1914), *Science and Method* (Nelson, reprinted by Dover).
PUGH, D. E. and Associates (1969), Facilities for the handicapped, *Architects' Journal*, March.
RESCHER, N. (1969), *Introduction to Value Theory* (Prentice-Hall).
ROSE, H. and S. ROSE (1970), *Science and Society* (Penguin).
ROSZAK, T. (1970), *The Making of a Counter-culture* (Faber and Faber).
ROY, R. (1971), The Future of Public Services, *Ph.D. Thesis*, University of Manchester Institute of Science and Technology.
RYAN, J. (1970), *Personal Communication to Stephen Platt.*
SAINSBURY, S. and P. TOWNSEND (in press), *The Disabled in Society: a study of persons registered as disabled by local authorities in London, Essex and Middlesex.*
SALIM, M. (1962), *The Marsh Arabs of the Euphrates Delta* (Athlone Press).
SARTRE, J.-P. (1965), *The Philosophy of Existentialism* (Philosophical Library).
SEFTWICH, R. H. (1966), *The Price System and Resource Allocation* (Holt, Rinehart and Winston).
SCHON, D. (1970), Dynamic Conservatism, *The Listener*, 26 November.
SCHÜTZENBERGER, M. (1954), A Tentative Classification of Goal-seeking Behaviours, *Journal of Mental Science*, *5*, 100.
SIDDALL, J. N. (in press), *Analytical Decision-Making in Design* (Prentice-Hall).
SIEBERT, J. and G. WILLS (eds.) (1970), *Marketing Research* (Penguin).
STRINGER, P. (1970), The Architect is a Man, *Architectural Design*, August.
SUTHERLAND, G. H. (1970), *M.Eng. Thesis*, McMaster University.
THORNLEY, J. (1969), *Guide to Manchester for the Disabled* (Manchester Youth and Community Service).
TOWNSEND, P. (1967), *The Disabled in Society* (Greater London Association for the Disabled).
TURNER, J. (1968), The Squatter Settlement: an architecture that works, *Architectural Design*, August.
UDDEN, P. (1966), *Technical Aids for the Handicapped: the Permobil, a crosscountry wheelchair* (Teknisk Hjålp åt Handikappade: Timrå, Sweden).
UMPLEBY, S. and J. BRIGGS (1970), Exploring the Future with a Computer, *The Futurist*, December.
WASKOW, A. I. (1969), Looking Forward 1999, in JUNGK, R. and J. GALTUNG (eds.), *Mankind 2000* (Allen and Unwin).
WILLIAMS, K. G. (1971a), Health and Society: the need for a new scientific attitude, *Community Health*, *2*, 4.
WILLIAMS, K. G. (1971b), Medicine, Engineering and Society, *Journal Inst. Mech. Eng.*, *18*, 6.
WILLIS, J. (1970), Information for Planning, *Architectural Design*, October.
WRIGHT, B. A. (1960), *Physical disability; a psychological approach* (Harper and Row).
YOUNG, P. T. (1968), Determinants of Belief, in BRITT, S. H. (ed.), *Consumer Behaviour and the Behavioural Sciences* (Wiley).